Constructibility
and
Mathematical Existence

Constructibility
and
Mathematical Existence

CHARLES S. CHIHARA

CLARENDON PRESS · OXFORD
1990

Oxford University Press, Walton Street, Oxford OX2 6DP
Oxford New York Toronto
Delhi Bombay Calcutta Madras Karachi
Petaling Jaya Singapore Hong Kong Tokyo
Nairobi Dar es Salaam Cape Town
Melbourne Auckland
and associated companies in
Berlin Ibadan

Oxford is a trade mark of Oxford University Press

Published in the United States
by Oxford University Press, New York

British Library Cataloguing in Publication Data
Chihara, Charles S.
Constructibility and mathematical existence.
1. Mathematics. Theories
I. Title
510′.1
ISBN 0-19-824817-2

Library of Congress Cataloging in Publication Data
Constructibility and mathematical existence / Charles S. Chihara.
Includes bibliographical references.
1. Mathematics—Philosophy. 2. Constructive mathematics.
3. Logic, Symbolic and mathematical. I. Title.
QAB.4.C45 1990   511—dc20   89-37162
ISBN 0-19-824817-2

Typeset by Cotswold Typesetting Limited, Gloucester
Printed in Great Britain by
Courier International Ltd., Tiptree, Essex

For
Michelle, ma belle,
qui trouve les mots qui vont très bien ensemble.

A philosophical problem is a call to provide an adequate explanation in terms of an acceptable basis. If we are ready to tolerate everything as understood, there is nothing left to explain; while if we sourly refuse to take anything, even tentatively, as clear, no explanation can be given.

Nelson Goodman, *Fact, Fiction, and Forecast*

PREFACE

THIS work is concerned with mathematical existence, a topic to which I have devoted a great deal of thought over many years. It might be regarded as a continuation of the study of mathematical existence found in my *Ontology and the Vicious-Circle Principle*; however, the predicative standpoint of that work has been abandoned in the present one. Why? The earlier position was developed when Quine's influence was especially strong. At that time, the free-wheeling use of modal notions seemed so fraught with problems that I restricted my modal mathematical developments to very limited constructions—hence the predicative approach. Quine's extensionalist views no longer hold such sway: few philosophers today believe that "true logic" should be restricted to first-order extensional logic. The person most responsible, I believe, for this shift in philosophical trends is Saul Kripke. His logical and philosophical writings on modality have led the way to a renewed interest in modal logic and to an invigorated use of modal notions in philosophical theorizing. The resurgence of research on modal reasoning came at an appropriate time for the development of the theory of constructibility that I was attempting to construct (see my (Simple) for an earlier version): it provided me with another basis for explicating, clarifying, and justifying some of the fundamental ideas of the theory.

This book has *not* been written only for researchers and scholars in the area of philosophy of mathematics. Those who are comfortable working with logical and mathematical concepts should find no difficulty in starting this book at the beginning and reading it straight through to the end; but others, especially those with only a general philosophical background, may find it preferable to proceed directly to Part II after reading only the first (introductory) chapter from Part I. They might then be motivated to work through the logical and mathematical material of Part I.

Serious writing of this book began in Paris in the fall of 1985. A sabbatical leave from the University of California and a National Endowment for the Humanities fellowship enabled me to devote the academic year to this project. Early versions of some of the chapters in this book were presented to scholarly groups at various times during that year. These include: the Conference on Scientific and Philosophical Controversy held at the Universidade de Évora in December 1985; the Logic Seminar held at L'Université René Descartes in January and April 1986; the Philosophy Seminar of the Universität des Saarlandes; and the Seminar of Philosophy and Mathematics held at the École Normale Supérieure in April 1986. The material I had written during this leave was then presented and discussed in a graduate seminar I gave in the fall of 1986; this enabled me to sharpen the ideas, to eliminate many unclarities and mistakes, and to develop the anti-Platonic position of the book. Another seminar, given in the fall of 1987, was devoted to some of the material of Part II. More recently, I have used material from this work in presentations I have given at the Philosophy Colloquium of the University of Washington, at a meeting of the Pacific Division of the American Philosophical Association, and at the Logic Colloquium of the University of California, Berkeley. The suggestions and criticisms I received from those who attended the above-mentioned presentations and seminars were very helpful to me.

Since the use of quotation marks in logical studies can be quite confusing to the uninitiated, a brief explanation of how quotation marks are used in this work may be in order. Double quotation marks are used for direct quotation and as "scare quotes" (i.e. as indicators that the expression appearing between the quotation marks, although accepted by certain people or groups, is not endorsed—or wholeheartedly endorsed—by the writer). Single quotation marks are used to refer to linguistic items (sentences, words, symbols, and the like): when a pair of single quotation marks is used, it is the linguistic item enclosed with the quotation marks that is being talked about. I also make use of the convention of displaying linguistic items in order to refer to them. For example, one can refer to the question 'Does Lois call Clark 'Clark'?' by displaying it, as it is done below:

Does Lois call Clark 'Clark'?

I use single quotation marks in the sentence displayed above because I understand it to be saying what is said by:

Does Lois call Clark by the name 'Clark'?

There will be cases in which one could use either single or double quotation marks with very little difference in sense. The sentence

Kitcher calls this system 'Mill Arithmetic'.

(in which I use single quotation marks as in the previous example) differs little from

Kitcher calls this system "Mill Arithmetic".

(in which I use double quotation marks to indicate that I am quoting Kitcher).

I wish to thank those students and former students, both graduates and undergraduates, who, through their participation in the above-mentioned seminars or in other courses I have taught, and through their criticisms, questions, and suggestions, have influenced the writing of this work. Among these, I should mention: Juan Bagaria, Craig Bach, Aleksandar Ignjatović, Ralph Kennedy, David Libert, Malcolm MacFail, Eugene Mills, Piers Rawlings, Phyllis Rooney, and Klaus Strelau.

Special mention should be made of Vann McGee, whose extensive comments on a draft of Part I were especially helpful. McGee and Jack Silver attended a seminar I gave in the fall of 1981 when I was still groping for a constructibility version of Simple Type Theory. The objections, critical comments, and suggestions these two astute logicians offered during the seminar greatly influenced the development of my views on this topic. It was McGee who first suggested to me that I use the possible worlds semantics to clarify the logic of the constructibility quantifiers.

The index was compiled by Susan Vineberg, who served as my research assistant for this project. She was a big help to me in getting the manuscript into publishable form. Susan carefully read several drafts of this work, checking them for errors, omissions, unclarities, inconsistencies, and the like. And she made many helpful suggestions, which I am sure have improved this work.

I wish to thank some of the philosophers whose writings I discuss in detail: Stuart Shapiro, Michael Resnik, John Burgess, Penelope Maddy, and Philip Kitcher have all sent me extensive comments on earlier versions of what I present in Part II. As a result, I am sure, my expositions of their views have become more accurate and my criticisms clearer; and for this, I am most grateful. I should also state, especially in view of my sharp disagreements with the above (and with Hartry Field), that I consider the writings I discuss in detail in Part II to be important works in the philosophy of mathematics, which are well deserving of careful study and analysis. I have learned much from these works.

Finally, it is with great pleasure that I acknowledge the aid I have received from my own family. My wife, Carol, has aided me in writing this book in countless ways, as only a wife can. But I owe her special thanks for the invaluable help and advice she gave me in dealing with the many computer and word-processor problems that arose in transferring the successive versions of this work on to disks. As for my daughter, Michelle, I have benefited much from discussing with her elementary mathematical ideas and problems, and also from following her mathematical education, both French and American, from her first pre-kindergarten classes to the present. I suspect that many views about the nature of mathematics would never have found their way into print had the authors studied close up the actual day-to-day process of learning mathematics that children go through.

Berkeley, California   C.S.C.
October 1988

ACKNOWLEDGEMENTS

I would like to thank the University of Chicago Press for permission to quote from Robert Geroch's *Mathematical Physics*; and Oxford University Press for permission to quote from Philip Kitcher's *The Nature of Mathematical Knowledge*. I would also like to thank the MIT Press for permission to produce (in Chapter 1) some material from my paper "Quine's Indeterminacy", which was published in *On Being and Saying* (edited by Judith Jarvis Thomson) copyright © 1987 by Massachusetts Institute of Technology. Pages 6–8 of the present work contain passages drawn from that paper.

CONTENTS

Part I
THE CONSTRUCTIBILITY THEORY

THE first part of this work begins with an introductory chapter in which a fundamental problem of philosophy of mathematics is presented. Several outstanding attempts to deal with this problem are discussed in this opening chapter. There follows a logical development of the basic ideas of the approach to be taken in this work. A formal deductive theory—the constructibility theory—is then set forth; and a justification of the axioms of this theory is given. Some objections to the system are considered and answered. Various mathematical developments—especially in cardinality theory and analysis—are carried out within the framework of the constructibility theory. Preliminary indications of the philosophical significance of the theory are supplied.

1

The Problem of Existence in Mathematics

1. The Philosophical Problem

AMONG the theorems of classical mathematics, one finds countless existence assertions. For example: 'There exists a real number which is such that its cube minus seven is zero' and 'There exists a set whose members are just those sets that have as members only finite cardinal numbers'. Now what is the significance of asserting that some mathematical entity (or some sort of mathematical entity) *exists*? And what does it mean? How, in short, are we to understand existence assertions in mathematics?

Philosophers might reasonably ignore such questions did they not also believe that *mathematics is true*, i.e. that the assertions of mathematics are for the most part true. Of course, not all people believe that mathematics is true. A Parisian mathematician once asserted, during a lecture I was giving, that in France children are not taught that the assertions of mathematics are true—instead, they are taught that these assertions are good![1]

It would seem, initially at least, that the path of least resistance is taken by those philosophers in what I shall call the Literalist Tradition. These philosophers agree with the majority of mankind that mathematics is true. They also maintain that the existence assertions of mathematics are not essentially different from the existence assertions of the empirical sciences, except, of course, the things asserted to exist are mathematical objects—not physical objects. One reason such a position may appear plausible is because, on the standard (and widely accepted) logical analysis of mathematical language, the existential assertions of mathematics are analysed in terms of the ordinary existential quantifiers of first-

---

[1] My daughter, who attended a lycée in Paris for a year, has assured me that this remarkable claim is not true.

order logic and are thus analysed to assert, in a literal and straightforward fashion, the actual existence of mathematical objects. In this work, I shall discuss some outstanding examples of philosophical views clearly in the Literalist Tradition: those of Willard Quine, Kurt Gödel, John Burgess, and Penelope Maddy (the last-two-mentioned views to be taken up in Chapters 9 and 10 respectively). Another position, Mathematical Structuralism (to be discussed in detail in Chapters 7 and 8), can be regarded as a modification of a Literalist view.

Philip Kitcher has put forward a contrasting anti-Literalist view, according to which what appear to be existential assertions in mathematics are in reality assertions to the effect that an ideal agent has performed certain operations. Kitcher's radical anti-Literalist views will be examined in detail in Chapter 11.

An even more radical view rejects the assumption that mathematics is true—at least in the straightforward way that mathematics is believed to be true by the Literalist philosophers. At one point, Hilary Putnam espoused such a view of mathematics. Making use of some ideas of the early Bertrand Russell, Putnam argued that "pure mathematics consists of assertions to the effect that *if* anything is a model for a certain system of axioms, *then* it has certain properties" (Thesis, p. 294). Although Putnam gave up this position not long afterwards, recently Hartry Field has advanced an instrumentalistic view of mathematics that was derived, in good measure, from Putnam's "If–thenism". I shall discuss Field's view in detail in Chapter 8. Another position (Arend Heyting's), to be discussed shortly, does not attempt to deal directly with this philosophical problem and undertakes instead to produce a kind of mathematics in which existence assertions of the sort that give rise to this problem are absent.

Let us explore briefly the Literalist's response to the problem of existence in mathematics. According to the Literalist, an assertion of existence in mathematics is not essentially different from an assertion of existence made in ordinary workaday contexts or in the empirical sciences: to say that a natural number greater than a million exists is just to say that there really exists a natural number which is greater than a million. Thus, to say that a set of such-and-such a sort exists is to say the same sort of thing one says in physics when one says that a molecule of such-and-such a sort exists: in the latter case, one is saying that *there is something* which is a molecule and which is of such-and-such a sort; whereas in the former case, one is saying that there is something which is a set and which is of such-and-such a sort. An obvious advantage to this kind of approach to the problem I have sketched above is that the "logic" of mathematical existence assertions is basically simple and well worked out. Mathematical existence assertions can be captured by the familiar existential quantifier of mathematical logic and there would be no need to distinguish the logic of mathematical existence assertions from that of any existential assertions of first-order logic.

Whatever the merits of such a response to the problem I have been sketching, it can be seen that the approach quickly leads us into some rather murky philosophical waters. The assertion 'It was discovered that there really are black swans' naturally gives rise to the question 'Where are these black swans to be found?', but the assertion 'It was discovered that there are continuous functions that are nowhere differentiable' gives rise to no analogous question. It seems to make no sense, in the case of mathematics, to ask 'Where are these functions?' Evidently, mathematical entities do not exist anywhere in physical space. Thus, the classical mathematician is regarded as asserting the existence of "abstract" entities not to be found in the physical world. The mathematician who believes the theorems of classical mathematics to be true, according to this position, ought to believe that there really are such entities as natural numbers, functions, sets, ordered pairs, and the like. The view that emerges then is that of the mathematician investigating a realm of entities that cannot be seen, felt, heard, smelled, or tasted, even with the most sophisticated instruments. But if this is so, how can the mathematician know that such things exist? We seem to be committing ourselves to an impossible situation in which a person has knowledge of the properties of some objects even though this person is completely cut off from any sort of causal interaction with these objects. And how does the mathematician discover the various properties and relationships of these entities that the theorems seem to describe? By what powers does the mathematician arrive at mathematical knowledge? In short, how is mathematical knowledge possible? As we shall soon see, those who espouse the kind of view under consideration attempt to provide plausible answers to these queries.

2. Quine's Platonism

How is knowledge of mathematical objects possible? For the philosopher-logician Willard Quine, the answer is straightforward: we obtain knowledge of mathematical objects in the same way that we obtain knowledge of physical objects, that is, through sense experience. Of course, we cannot literally see or observe the objects typically studied in mathematics courses, but we cannot see or observe the objects typically studied in quantum physics courses either. In both cases, it is claimed, we can obtain evidence supporting the postulation of the objects we investigate.

Sets, for Quine, are unchanging abstract objects the existence of which it is reasonable to postulate. For we have the same sort of reasons for postulating the existence of sets, he believes, that we have for postulating the existence of molecules: basically, we have a kind of scientific evidence for the existence of sets.[2]

To understand Quine's views on this matter, it is useful to review the principal points in his (Posits). In this article, Quine considers the question of what evidence the scientist can muster for belief in the existence of molecules, given that molecules are too small to be directly observed. The answer generally given is that there is a large fund of "indirect evidence" obtained by experimentation and theory testing involving a variety of different phenomena from such diverse areas of study as biology, chemistry, and atomic physics. For Quine,

The point is that these miscellaneous phenomena can, if we assume the molecular theory, be marshalled under the familiar laws of motion. The fancifulness of thus assuming a substructure of moving particles of imperceptible size is offset by a gain in naturalness and scope on the part of the aggregate laws of physics (Posits, pp. 233–4).

Thus, the "indirect evidence" mentioned above turns out to consist in certain "benefits" engendered by the acceptance of the doctrine of molecules. These "benefits" turn out to be: simplicity of theory, familiarity of the principles used, increase in the scope of testable consequences, extension of theory, and agreement with experimentation. But these are "benefits" that give the physicist reasons for using the molecular theory. Do they also afford any real evidence of

---

[2] The Quinian position I describe here was developed in the 1950s and 1960s (W and O being the central work of this period). Quine's more recent writings suggest that there have been some subtle changes in his Platonic views.

the truth of the molecular theory? "Might the molecular doctrine", Quine asks, "not be ever so useful in organizing and extending our knowledge of the behavior of observable things, and yet be factually false?" (Posits, p. 235). He does not attempt to answer this question straight off but instead investigates the reasons we have for believing in common-sense bodies, such as tables and automobiles. After all, he suggests, what we are directly aware of are such things as visual patches, tactual feels, sounds, tastes, and smells. So if we have any evidence for the existence of common-sense objects, Quine claims that "we have it only in the way in which we may be said to have evidence for the existence of molecules. The positing of either sort of body is good science in so far merely as it helps us formulate our laws" (p. 237).

Quine then reconsiders the question of whether such "benefits" as simplicity of theory and familiarity of principles that are garnered through the postulation of molecules provide us with any real evidence of the truth of the molecular theory. He concludes that if we take the position, in the molecular case, that no evidence is provided, then we would be stuck with holding that we have no evidence for the existence of common-sense bodies either. We should have to say that tables and automobiles are unreal. But this, according to Quine, would be absurd: "Something went wrong with our standard of reality. . . . Unless we change meanings in midstream, the familiar bodies around us are as real as can be; and it smacks of a contradiction in terms to conclude otherwise" (Posits, p. 238). The absurdity of concluding that bodies are unreal can also be seen from the fact that such a conclusion vitiates the very considerations that led us to it, claims Quine: To suppose that we have posited the existence of bodies in order to systematize the data of our senses is absurd, for past sense-data are lost except for what is stored in our memory, "and memory, far from being a straightforward register of past sense data, usually depends on past posits of physical objects" (p. 238). So we must conclude, Quine asserts, both that *the "benefits" we mentioned above do count as evidence*, indeed "the best evidence of reality we can ask" and also that "we can hope for no surer touchstone of reality".

Such being what evidence is, according to Quine's lights, we can understand his response to the logical positivist's position that one cannot have empirical evidence for such statements as 'sets exist':

I think the positivists were mistaken when they despaired of evidence in such cases and accordingly tried to draw up boundaries that would exclude such sentences as meaningless. Existence statements in this philosophical vein do admit of evidence, in the sense that we can have reasons, and essentially scientific reasons, for including numbers or classes or the like in the range of values of our variables (Exist, p. 97).

What are these "scientific reasons"? They turn out to involve very general considerations of the sort described in Quine's discussion of physical bodies and molecules. Indeed, at one point, Quine suggests comparing the "benefits" of postulating attributes with those resulting from the postulation of classes (Speak, p. 23). Thus, problems with individuation of attributes might require a certain disruption of the logic of the overall conceptual scheme, and this might require a more complicated logical theory than would be required if only classes were postulated. In this way, simplicity considerations would indicate that classes should be postulated instead of attributes. What is clear is that the weighing of the kind of evidence talked about in this context is a rather nebulous and vaguely specified matter, which involves comparisons of the overall theories that result from adopting competing hypotheses.

In the above quotation, Quine speaks of including such things as classes and numbers in the range of values of our variables. How are we to understand such talk? Basically, Quine imagines that he is in the process of combining all of our scientific theories into one grand theory, which is to be formulated in a first-order logical language. (Why Quine chose first-order logic as the logic of science will be taken up shortly.) He then asks himself: what sorts of things would have to be in the range of values of the variables in order for the theory to be true? Well, presumably there would have to be various kinds of physical objects, such as molecules and tables, in the range of values of the variables. But what about mathematical objects? He writes:

Certain things we want to say in science may compel us to admit into the range of values of the variables of quantification not only physical objects but also classes and relations of them; also numbers, functions, and other objects of pure mathematics. For mathematics—not uninterpreted mathematics, but genuine set theory, logic, number theory, algebra of real and complex numbers, . . . —is best looked upon as an integral part of science, on a par with physics, economics, etc., in which mathematics is said to receive its applications (Scope, p. 231).

Now this quotation suggests that Quine should believe in the existence of all sorts of mathematical objects: numbers, functions, Hilbert spaces, ordered pairs, etc. But in fact, Quine holds that we should postulate only the existence of classes. The reason? Ontological economy: "Researches in the foundations of mathematics have made it clear that all of mathematics in the above sense can be got down to logic and set theory, and that the objects needed for mathematics in the above sense can be got down to a single category, that of *classes*" (Scope, p. 231). The idea seems to be that in this one grand theory all the mathematics we shall need will be set theory, since all the other branches of mathematics needed in science can be reproduced in set theory. Thus, the only mathematical objects that are required in this theory will be sets.

Why, it might be asked, *should* this grand theory be formulated in first-order logic? I sometimes wonder if Quine's reasons for his choice of such a logical language do not come down to simply a case of wishful thinking ("Wouldn't it be nice if . . . ?") However, Quine has at various times given various reasons for selecting first-order logic as the language of science, and we should consider these. One sort of reason he gives is the clarification and simplification of the "conceptual scheme of science" that is obtained by this choice: "Each reduction that we make in the variety of constituent constructions needed in building the sentences of science is a simplification in the structure of the inclusive conceptual scheme of science" (W and O, p. 161). He goes on to say: "The same motives that impel a scientist to seek ever simpler and clearer theories adequate to the subject matter of their special sciences are motives for simplification and clarification of the broader framework shared by all the sciences" (p. 161). Settling on first-order logic eliminates various sorts of confusions and ambiguities that could arise in scientific discourse and it also promotes "programming deductive techniques" (p. 231). But Quine has another sort of reason for his choice: his famous *criterion of ontological commitment* is given in terms of first-order languages and he believes that only by formulating scientific theories in such a language can the ontological commitments of science be made clear.[3]

A full-scale critique of Quine's reasoning would be out of place here, but a few critical remarks are in order. The argumentation in (Posits) is questionable at several points. The crucial step in Quine's deliberations occurs after he has concluded that it would be absurd to

---

[3] For a detailed discussion of this criterion, see my (V-C), ch. 3, sect. 3.

suppose that physical bodies are unreal: "Having noted that man has no evidence for the existence of bodies beyond the fact that their assumption helps him organize experience, we should have done well, instead of disclaiming evidence for the existence of bodies, to conclude: such, then, at bottom, is what evidence is, both for ordinary bodies and for molecules" (Posits, p. 238). As Quine sees it, unless we grant that the pragmatic benefits generated by the assumption of bodies do count as *evidence* for the existence of bodies, we shall be forced to conclude that bodies are "unreal". But, Quine reasons, such a conclusion would be absurd. So we must grant that these pragmatic benefits do count as evidence.

In this argument, Quine assumes that *if we conclude that there is no evidence for the existence of physical objects, we would have to conclude also that physical objects are unreal*. But that is not at all obvious. The following position, it seems to me, can be given a plausible defence: Our belief in the existence of physical objects is so fundamental to our thinking and theorizing that all talk of evidence for such a belief is inappropriate. The practice of gathering evidence, confirming hypotheses, testing theories, etc. takes place within a framework of ideas in which physical objects are presupposed. But that does not mean that physical objects should be taken to be unreal. It is not that our belief in physical objects is like the belief that some people may have in gremlins. In the case of gremlins, it is appropriate to ask for evidence for the belief and to try to gather evidence for it. It just so happens that evidence is lacking. Our belief in physical objects, however, may be just too fundamental to our thinking for the question of evidence for the belief to be appropriate. Thus, Quine's inference from the absurdity of supposing that physical objects are unreal to the conclusion that we must have evidence for the existence of physical objects is not entirely compelling. We can thus see that Quine has not yet produced a very convincing argument for his position that such "benefits" as simplicity and familiarity of principle, attributable to the postulation of some type of object, *must* count as evidence for the existence of this type of object.

But even granting Quine his views on evidence, his argument that we have scientific evidence for the existence of classes can be seen to be defective. Notice that Quine does not seriously undertake the task of formulating science as a first-order theory. Not a single area of a genuine empirical science is given the required treatment. It may appear to some philosophers or logicians that constructing such a first-order theory would be simply a matter of carrying out a rather mundane, practically difficult, but theoretically empty task. So it is useful to ponder just how one would go about constructing such a theory. Would one take up various physics journals and textbooks and simply translate what one found there into a first-order language in the way one does in beginning logic courses? Apart from the well-known problems of translating English prose into logic, there are particular problems connected with the nature of physical theories and laws. Even such a fundamental law as Coulomb's Law, a version of which states that every point charge produces a surrounding spherically symmetric electromagnetic field, raises serious problems: straightforwardly translating Coulomb's Law as a first-order sentence results in something that is clearly false. For, as Geoffrey Joseph has noted, if strong gravitational forces are acting on the point charge, it may well not be surrounded by a spherically symmetric electromagnetic field:

> The general-relativistic theory of the gravitational field informs us that if one places a very dense source of mass-energy near the point charge, the metric of space-time is changed in that vicinity. The quanta of the electromagnetic field, having nonzero mass-energy, respond to this gravitational influence. Their resultant trajectories are thus altered from the spherically symmetric distribution (Many, p. 776).

Thus, in order to have a version of Coulomb's Law that is true, it is frequently suggested that it should be understood to involve a *ceteris paribus* or, better yet, a *ceteris absentibus* clause: what it should be taken to say, essentially, is that every point charge *would produce* a surrounding spherically symmetric electromagnetic field, were all other interfering forces absent. The idea is that what is described by Coulomb's Law are ideal situations, or situations in which complicating forces are absent. But notice, one has changed the statement of the law from the indicative to the subjunctive mode (Many, p. 777), so it becomes questionable whether a translation of the law into the extensionalist's first-order language is even feasible. Furthermore, the law will no longer apply in a direct straightforward way to most real situations, since most situations we meet in real life involve complicating forces (see Cartwright (Lie), Essay 2). Of course, this particular problem about translating Coulomb's Law into first-order logic is just one of many that a first-order extensionalist would have in dealing with field theories. It can be seen that what would be required here is scientific theory construction and sophisticated analysis of scientific practice, going far beyond simple translation. All the hard theoretical work that would be involved in such an undertaking is bypassed by Quine's procedure. Quine never goes about producing the first-order theory in question; he just imagines that he is doing it. Needless to say, it is difficult to compare this pretend theory with actual ones.

Right off, then, we see something very questionable in Quine's reasoning. The supposed "scientific reasons" we have for positing and believing in sets involves imagining that we are constructing a first-order version of all of science and then considering what sorts of entities we would have to refer to in such a theory. Since none of the details is supplied, we have only the vaguest of ideas of what such a theory would be like. In this context, we are supposed to decide:

(*a*) what sort of mathematical theory we would include in the theory;

and

(*b*) whether the theorems of this mathematical theory should be treated as true statements on a par with those of the other sciences (physics, biology, etc.), or should be regarded in some other way.

But these decisions are to be made on the basis of an evaluation of the "scientific benefits" of the alternatives considered. Furthermore, one is to assess these "benefits" in terms of the resulting wholes, i.e. one is to compare the whole of the scientific theory one would get if one chose one alternative with the whole of the scientific theory one would get if one chose another. Thus, in discussing *simplicity*, one of the principal "benefits" to be gained by positing molecules, he says: "There is a premium on simplicity in the hypotheses, but the highest premium is on simplicity in the giant joint hypothesis that is science, or the particular science, as a whole. We cheerfully sacrifice simplicity of a part for greater simplicity of the whole when we see a way of doing it" (Web, p. 45). One might wonder what confidence we should have in any evaluations made in this context, when one is not considering any actual scientific theories, but only the faint outlines of imagined ones obtained by large and untested extrapolations from relatively simple aspects of a few scientific theories. Indeed, it is worth asking why one should believe that such a first-order theory would be a scientifically acceptable alternative to what we now have. Quine has emphasized certain benefits that would flow from the accomplishments of this task. These are benefits that a philosopher-logician would think of: there would be clarification and simplification of the conceptual scheme of science (W and O, sect. 33), and the logic of the system would be the familiar and thoroughly worked out first-order logic. In addition, the ontology and the ontological commitments of science would be made definite. Let us grant, for the sake of argument, that there would be these benefits. But none of these benefits would be decisive. We have got by in science and mathematics for over two thousand years with theories formulated in good old-fashioned natural languages, without insuperable difficulties due to ambiguities or to the lack of a complete logical proof procedure. And Quine's suggestion (in Facts, p. 161) that anyone who is unwilling to formulate her theory in the language of the predicate calculus should be regarded as being unwilling to reveal the ontological implications of the theory is surely questionable: serious scientific disputes about the ontological implications of our scientific theories—theories formulated in natural languages—have been (and are being) carried out. (I shall say more about this point in Chapter 3, Section 6.)

Besides, what about the more clearly *scientific* benefits—benefits that an empirical scientist would find telling? In the absence of more substantial scientific benefits, might it not be that no first-order formulation of "all of science" (whatever that would be) would be as good scientifically as what we now have? Besides, we clearly need to do more than just look at the advantages that would flow from formulating science in a first-order language; in order to decide whether such a reformulation would be justified, we need also to weigh the *disadvantages*, both theoretical and practical, that would result from such a formulation. And this, so far as I can see, Quine has never done. What we find by way of justification is the mere citation of a few advantages, and never any attempt to make an overall assessment which takes into account the enormous practical problems involved in such a reformulation of science. One need only think of the radical changes that would be required in textbooks, in scientific education, and in formulation of theory to get some small idea of what would be involved.

It might be thought that many of the disadvantages I have mentioned in the previous paragraph are merely pragmatic (as opposed to theoretic) ones, and hence should not be considered in assessing the sort of theoretical question under discussion. But for Quine, assessing the acceptability of the theoretical hypotheses of the empirical sciences is a pragmatic matter: he once went so far as to suggest (in Carnap, p. 125) that one's hypotheses about what there is, as for example whether molecules or sets exist, "is at bottom just as arbitrary or pragmatic a matter as one's adoption of a new brand of set theory or even a new system of bookkeeping". And in (Dogmas), he espoused a thoroughgoing pragmatism according to which the only rational considerations that determine what scientific hypotheses a person ought to believe are pragmatic ones (p. 46).

Michael Resnik once responded (personal communication) in the following way to the above objection to Quine's views: Quine does not claim, said Resnik, that scientists should actually use his first-order canonical language; Quine recognizes that the purposes of the philosopher of science differ from those of the working scientist. I agree. But this response is based on a misunderstanding of the above criticism. I am not objecting that Quine's ontological claims are implausible because it would be impractical or inconvenient for practising scientists to use the Quinian logical language. Rather, I am questioning Quine's claim to have scientific evidence to support his belief in the existence of sets: I am arguing that, even using Quine's own pragmatic standards of evidence, Quine has not shown that his ontological hypothesis is supported.

Re-examine the quotation from (Exist) I gave earlier in which it is claimed that we have evidence for the existence of sets. It is suggested that this evidence consists in "the power and facility" which the positing of sets contributes to physics. But as we have seen above, Quine's own views of evidence require a judgement about how the postulation will affect the whole of science. (This is especially the case when the postulations involve something so fundamental to scientific theory as do those of mathematics.) So it is not enough to cite some benefits to overall theory that may be derived from the postulations: we need to see what effects this postulation would have on the whole picture, and in the process, to weigh not just the advantages that flow from the postulation, but also (as I noted above) the disadvantages. And this whole picture involves not just the traditional areas of science, but also the more philosophical ones, such as theory of knowledge and metaphysics. What effect the positing of abstract mathematical objects would have on the epistemological part of this grand science is difficult to assess. In the absence of an actual first-order theory which is to be affected by such a postulation, what confidence can we have in Quine's speculations? Do we have evidence here, in any real sense?

That positing mathematical objects is to be justified in the same way that physicists justified their positing of molecules is intimately linked to another idea that emerges from Quine's writings on this topic: it is suggested that which mathematical theory we should take to be true should be determined empirically by assessing the relative scientific benefits that would accrue to science from incorporating the mathematical theories in question into scientific theory. It is as if the mathematician should ask the physicist which set theory is the true one! I do not believe that many mathematicians find this Quinian picture very appealing.

3. Gödel's Platonism

The great logician Kurt Gödel has suggested an answer to the question 'How is mathematical knowledge possible?' It is, in a word: intuition, i.e. mathematical intuition. Just as sense experience is the means by which we gain knowledge of physical objects in physics, mathematical intuition serves as the means by which we gain knowledge of mathematical objects in mathematics: "I don't see any reason why we should have less confidence in this kind of perception, i.e., in mathematical intuition, than in sense perception, which induces us to build up physical theories and to expect that future sense perceptions will agree with them" (Cantor's, p. 271). However, to understand Gödel's views on this matter, one must investigate his reasons for believing in mathematical objects.

Gödel maintained that the question of the existence of mathematical objects "is an exact replica of the question of the objective existence of the outer world" (Cantor's, p. 220). For Gödel, to ask if mathematical objects exist is to ask a question entirely analogous to asking if physical objects exist. Both questions, he believed, should be answered in the affirmative. And he believed in the existence of mathematical objects for reasons related to those given by Quine. Gödel believed, as did Quine, that we assume or postulate the existence of physical objects in order to organize and have a satisfactory theory of our sense experience. And he thought we should assume the existence of such mathematical objects as sets for analogous reasons. According to Gödel, such an assumption "is quite as legitimate as the assumption of physical bodies and there is quite as much reason to believe in their existence" (Russell's, p. 220).

To see what lies behind these views regarding the existence of mathematical objects, consider a passage that occurs in his (Cantor's). The passage is to be found in the Supplement to the second edition, where he focuses on the possibility that the Continuum Hypothesis is undecidable from the accepted axioms of set theory. He contrasts this set-theoretical case of independence with the independence of Euclid's fifth postulate. In geometry, when the primitive terms are interpreted, the meaning is given in terms of physical objects and physical phenomena. But the set-theory case is different, we are told:

[T]he objects of transfinite set theory ... clearly do not belong to the physical world and even their indirect connection with physical experience is very loose (owing primarily to the fact that set-theoretical concepts play only a minor role in the physical theories of today).

But, despite their remoteness from sense experience, we do have something like a perception also of the objects of set theory, as is seen from the fact that the axioms force themselves upon us as being true. I don't see any reason why we should have less confidence in this kind of perception, i.e., in mathematical intuition, than in sense perception, which induces us to build up physical theories and to expect that future sense perceptions will agree with them and moreover, to believe that a question not decidable now has meaning and may be decided in the future (Cantor's, p. 271).

In context, it is clear that when Gödel speaks of "their remoteness from sense experience", he is referring to the *objects of transfinite set theory* (which, after all, are the objects relevant to the question of the truth of the Continuum Hypothesis); that is, he is referring to objects that do not belong to the physical world. That, among other reasons, is why these objects are remote from sense experience. Thus, although we cannot literally perceive these non-spatial, non-temporal objects, we are said to have *something like a perception of them*. And Gödel goes further: he implies that we can see that we have something like a perception of these objects from the fact that the axioms of transfinite set theory force themselves upon us as being true. In other words, what is put forward as justifying the claim that we have something like a perception of these objects is the fact that the axioms force themselves upon us as being true.

The crucial elements in Gödel's reasoning can now be set down as these: First of all, we have the idea, discussed in connection with

Quine's reasoning, that physical objects are assumed or posited to exist in order to organize, systematize, and have a workable theory of our sense perception. Second, it is claimed that we have something in the realm of mathematics that is like sense perception and that plays the role in the mathematical realm that sense perception plays in the physical realm, namely mathematical intuition. Third, there is the suggestion that we have the same kind of reason for postulating the existence of mathematical objects that we have for postulating the existence of physical objects, namely we need to postulate mathematical objects in order to organize and have a satisfactory theory of our mathematical intuitions—thus the claim that mathematical objects are "in the same sense necessary to obtain a satisfactory system of mathematics as physical objects are necessary for a satisfactory theory of our sense perceptions" (Russell's, p. 221). One can see why Gödel would think that the question of the existence of mathematical objects is entirely analogous to the question of the existence of physical objects. One can also see several respects in which Gödel's reasoning is close to Quine's. However, there are some striking respects in which Gödel's position is very different from Quine's. First, Quine held that we ought to postulate mathematical objects in order to have an adequate *scientific theory*, that is to say, both physical objects and mathematical objects, according to Quine, must be postulated to have a workable scientific theory that fits in with our sense experience. There is no appeal in Quine's reasoning to mathematical intuition, nor is there any suggestion that we need to have a theory that organizes and systematizes our mathematical intuitions in the way that our physical theories organize and systematize our sense perceptions. Second, Gödel has, in effect, provided an answer to the query 'How do we gain knowledge of the properties of these postulated objects that supposedly do not exist in the physical world?'—an answer that is very different from Quine's. For Gödel, we are able to perceive, in some sense, the objects in question. No such ability is claimed by Quine.

For a variety of reasons, I do not find Gödel's reasoning on this matter very convincing. Since I have given detailed critical evaluations of his position elsewhere,[4] I shall restrict myself here to disputing only one aspect of Gödel's views on this matter. Gödel maintains that we have a kind of perception of the objects of

---

[4] See my (V-C), ch. 2 and also my (Gödelian).

transfinite set theory, and this position is supposed to follow from the fact that the axioms of set theory force themselves upon us as being true. But how do the axioms force themselves upon us as being true? Take a specific example: consider the axiom of Zermelo–Fraenkel set theory known as 'the regularity axiom', which says that if a set x has an element, then it has a minimal element y, i.e. an element y such that x and y are disjoint. Presumably, this axiom forces itself upon one as being true, in contradistinction to the Continuum Hypothesis, which does not. However, this axiom will not force itself upon just anyone as being true: a person who knows nothing about sets, has no concept of set at all, will feel no more compulsion to accept the regularity axiom than he/she will to accept the Continuum Hypothesis. Furthermore, someone who learned about sets only through studying Quine's (NF) might not see any good reason for accepting the regularity axiom. Only someone who has an appropriate notion of set will feel such a compulsion. So consider the description of the intuitive notion of set described by Joseph Schoenfield in his well-known textbook:

We start with certain objects which are not sets and do not involve sets in their construction. We call these objects *urelements*. We then form sets in successive stages. At each stage we have available the urelements and the sets formed at earlier stages; and we form into sets all collections of these objects. A collection is to be a set only if it is formed at some stage in this construction (Logic, p. 239).

Then we can be "forced to accept the regularity axiom" by the following chain of reasoning sketched by Schoenfield: "To see that the regularity axiom is true, let x be a nonempty set, and let y be a member of x formed at as early a stage as possible. Since the members of y must be formed at a still earlier stage, they are not members of x. Hence y is a minimal element of x" (ML, p. 238). One can imagine a student reading this passage and then exclaiming: "Of course, the regularity axiom must be true!" But should we then say that the student must, in some sense, be able to perceive sets? Surely we cannot deduce any such thing: Gödel's inference is clearly not a purely logical inference. Evidently, Gödel's reasoning is based on the idea that the postulation of something analogous to perception for the case of mathematical objects best explains or accounts for this phenomenon of the axioms of set theory forcing themselves upon us as being true.

But there are problems with such reasoning. First of all, the appeal to a kind of perception of mathematical objects does not seem to yield anything like a satisfactory explanation of the phenomenon under consideration. After all, there is supplied no description of a causal mechanism by which we humans are able to "perceive" objects that do not exist in the physical world. The appeal to mathematical intuition does not explain how we are able to "perceive" sets—it, essentially, only asserts that we do.

Secondly, it is by no means clear that a naturalistic explanation cannot be given of the phenomenon in question. For example, might not cognitive psychologists someday have a sufficient grip on the phenomenon to be able to offer some kind of reasonable explanation of it? It seems to be much more likely that we will eventually be able to give some kind of scientific explanation of this phenomenon than that Platonists will be able to fill the gaps in their metaphysical explanation.

However, there is another way of interpreting the quoted passage, which some may regard as more plausible than the above. What needs to be explained, according to this interpretation, is not the *psychological phenomenon* of the axioms forcing themselves upon us, but rather the epistemological state that obtains, that is, what Gödel proposes to explain by postulating a kind of perception of the objects of set theory is the fact that we come to have *knowledge* of the truth of the axioms of set theory. Crudely put, the suggestion is that the only plausible explanation of this knowledge is that we have a kind of perception of sets.

Certainly, if we assume that as a result of having the axioms of, say, Zermelo–Fraenkel set theory forced upon her, the student comes to have knowledge of objects that do not exist in the physical world, then the idea that we have some kind of perception of these objects has a certain attraction. But why should we assume that? Reconsider the regularity axiom. Does the student believe, as a result of her insight, that she has learned something about some objects that truly exist? After all, notice that the student could very well also learn that in a different set theory the regularity axiom does not hold. There would be no contradiction in her accepting the regularity axiom in one system of set theory, while also rejecting regularity in another set theory. This suggests that the knowledge in question is not knowledge of the properties of a truly existing realm of objects, but is only conceptual knowledge. Thus, when the student became convinced of the "truth" of the regularity axiom, what did she learn? I suggest two ways of regarding this. One can regard the student as having learned a sort of hypothetical fact: If there are sets and if the universe of sets has the structure given by the informal account provided by Schoenfield, then the regularity axiom will hold of this structure. We can also regard the student as having learned that regularity is *true to* the iterative conception of sets.[5] That is, the student learns that the conception of sets she has been given is such that regularity is implied. But the student does not learn that regularity holds of an actually existing universe. For how would she know that the universe of sets described in the informal account truly exists? After all, Schoenfield does not prove that there is such a universe of sets. Indeed, he does not even give any grounds for believing that there is any such universe. Instead, he talks about what is the case *according to a certain notion of set* (ML, p. 238). Thus, what is suggested is that the student learns that regularity must hold, given the iterative notion of set.

The kind of phenomenon under discussion of axioms forcing themselves upon us does not seem to be restricted to mathematics. One can think of similar things arising when someone is acquiring practically any new concept. For example, imagine a student acquiring an intuitive notion of 'perfect beings', it being explained to her that a perfect being is *something that has all perfections to the highest possible degree*. And suppose further that the student's notion of perfection is such that being necessarily existent, i.e. *not depending for one's existence on anything else*, is a perfection. Then the student could be forced to conclude that a perfect being will have the perfection of being necessarily existent. It would seem that the "axiom" that says 'All perfect beings have the perfection of being necessarily existent' forces itself upon her as being true. But surely we would not conclude from this that the student must have some kind of perception of perfect beings, or that any such perfect beings in fact exist. As in the set-theoretical case, what the student comes to know about perfect beings is hypothetical in nature: whatever is a perfect being (if there are any such things) must have the perfection of necessary existence. And this knowledge does not seem to presuppose that any perfect beings in fact exist. Gödel's justification for his claim that we must have a kind of perception of mathematical objects is simply not convincing.[6]

[5] See my (V-C), ch. 2, sect. 2, for a much fuller discussion of this notion of a statement being *true to a concept*.

[6] See my (V-C), ch. 2, and (Gödelian) for additional discussions of these matters.

In response to the above objection, it has been suggested that the set-theory case is different from the perfect-being example inasmuch as set theory (or some roughly equivalent mathematical theory) has proven to be an indispensable part of the empirical sciences. The theory of perfect beings described above is scientifically useless, whereas set theory is not. Doesn't this suggest that the axioms of set theory are linked to truth in a way that the axioms of the theory of perfect beings is not? One can see in this response, an echo of Quine's reasons for believing in the existence of sets, but without the trappings of reconstructing science in first-order logic. I shall leave for now the question of whether the usefulness of set theory in science, coupled with the fact that the axioms of set theory force themselves upon us as being true, gives us reasons for believing that we have some kind of perception of sets. This question will be taken up later in Chapter 12.

Whether or not one finds Gödel's reasoning concerning mathematical objects to be at all plausible, even supporters of the Gödelian view must admit that there are features of the view that make it difficult to accept. The mathematician is pictured as theorizing about objects that do not exist in physical space. This makes it appear that mathematics is a very speculative undertaking, not very different from traditional metaphysics. A mysterious faculty is postulated to explain how we can have knowledge of these objects. Gödel's appeal to mathematical perceptions to justify his belief in sets is strikingly similar to the appeal to mystical experiences that some philosophers have made to justify their belief in God. Mathematics begins to look like a kind of theology. It is not surprising that other approaches to the problem of existence in mathematics have been tried.[7]

4. Heyting's Intuitionism

One alternative to the Platonic views discussed above has been explored by the well-known Intuitionist Arend Heyting, who views the existence assertions of classical mathematics (as well as the classical idea of existence) as being somehow "metaphysical" in nature and *lacking in clear sense* (Int, pp. 2 and 11). Heyting's radical

---

[7] Penelope Maddy gives in (Perception) a very different interpretation of Gödel's Platonism than the one presented above. See my (Gödelian) for a criticism of her interpretation of Gödel.

solution to the above philosophical problems is to banish such existence assertions from mathematics altogether and to advocate the construction of a new mathematics along the lines set down by L. E. J. Brouwer. One of the principal ideas of Heyting's programme is to give the existential quantifier of Intuitionistic logic a quite different sense from that of classical logic. Heyting does not attempt to explain the existential quantifier of his system in terms of truth conditions, as is done in classical logic, but rather in terms of assertability conditions. Roughly, one is allowed in Heyting's logical system to assert a proposition $(\exists x)Fx$ just in case one can specify a mathematical object a which is such that one can assert that Fa. Unfortunately, the mathematical theories that the Intuitionists have developed, on this basis, are much weaker than the classical theories they have rejected. And few seem to believe that Intuitionistic mathematics is adequate for the needs of the empirical sciences (cf. (Int), p. 10). This is one reason I do not regard Heyting's Intuitionism as a plausible way of dealing with the problem of existence in mathematics.

In connection with the point just made, I would like to put forward for discussion a thesis that will undoubtedly strike many as being outrageous. I do so, not because I think the thesis is completely defensible, but because I think it is enlightening to do so. The thesis is this: *Heyting's Intuitionism is not so much a philosophy of mathematics as it is the construction of a new kind of mathematics.* Now how might this thesis be defended? More specifically, how might one defend the idea that Intuitionism, as Heyting has articulated it, is not really a philosophy of mathematics? We can imagine someone arguing as follows:

Philosophy of mathematics is the value of the function *philosophy of x*, for argument $x =$ mathematics. Consider values of that function for other arguments. Consider philosophy of art, of religion, of history. Do philosophers of religion try to produce a new kind of religion, or do they try to analyse and understand religion? Do philosophers of music try to produce a new kind of music, or do they try to provide philosophical insights into the nature of music? Take philosophy of science. Philosophers of science do not attempt to produce a new kind of science: we do not have Popperian science, Kuhnian science, Feyerabendian science, or Putnamian science. Instead, philosophers of science attempt to characterize the nature of science in a perspicuous and enlightening way; to uncover the basic principles according to which scientists operate; and to make our scientific theories and practices consistent with other areas of human understanding, and in the process, resolve (or dissolve) paradoxes and conceptual or methodological conflicts between scientific practice and theory, on the one hand, and common-sensical or philosophical beliefs on the other. In short, philosophy of science is directed at understanding science—and by 'science' I mean actual science, i.e. science as it is actually practised. Similarly, philosophy of mathematics is, and should be, directed at understanding actual mathematics, and dealing with the philosophical problems arising out of the enormous role (actual) mathematics plays in everyday life and in science. When Heyting turned his back on classical mathematics and gave up trying to understand classical mathematical assertions, he in effect gave up philosophy of mathematics!

As I said above, I do not consider the above argument completely defensible. The point of the objection is to show, in an exaggerated way, what little light is shed on the problem of existence by Heyting's procedure.[8] However, despite my misgivings about Heyting's views, my own approach to these problems is similar, at least in one respect, to what Heyting advocated.[9] If we are puzzled about certain aspects of classical mathematics, why not construct another kind of mathematics that will avoid those features of the original system that gave rise to the puzzles? A study of these alternative mathematical theories will give us a new perspective from which to view classical mathematics, which could prove to be extremely enlightening. This strategy may not give us a direct solution to the problem described in this chapter, but it may enable us to gain insights into the role classical mathematics plays when it is applied.

[8] Cf. Heyting's Int's comment: "Let those who come after me wonder why I built up these mental constructions and how they can be interpreted in some philosophy; I am content to build them in the conviction that in some way they will contribute to the clarification of human thought" (Int, p. 12).

[9] I do not wish to suggest by the above comment that I shall not be advancing in what follows any kind of philosophy of mathematics.

2

The Constructibility Quantifiers

1. Introduction

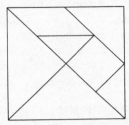

Fig. 1.

THERE is a Chinese puzzle consisting of a square cut up into five triangles, a square, and a rhomboid, these seven pieces of which can be combined so as to form a large variety of other figures. One is supposed to form these figures using all seven pieces and without distorting the shape of any of the pieces, say, by cutting or filing. Any figure formed by combining these seven pieces will be called here a '*tangram*'. Now it is possible to construct a tangram that is a triangle; it is also possible to construct a tangram that is a rectangle. Indeed, it is possible to construct tangrams in all sorts of different shapes. For example, it is possible to construct tangrams in the shape of letters: 'W', 'T', and 'L', are just a few.

I can imagine someone developing a theory about what tangrams it is possible to construct. Some of the theorems of this theory would assert the possibility of constructing tangrams of various sorts. And some of these constructibility theorems would be justified by showing how the seven pieces could be placed so as to form the tangram in question. The phrase 'it is possible to construct a tangram such that' functions very much like a quantifier. Indeed, it will be shown in this work how one can interpret the existential quantifier of a version of classical mathematics to be functioning very much like the constructibility phrase mentioned above. Notice that an assertion of the constructibility of a tangram is not an assertion that a tangram of the sort described actually exists; nor does it imply that anyone has actually constructed such a tangram. To say that a tangram of a particular sort is constructible is to say that it is possible to construct such a tangram. Whether a tangram of this sort actually exists is another question.

The basic idea of the approach to be taken in this work is to develop a mathematical system in which the existential theorems of traditional mathematics have been replaced by constructibility theorems: where, in traditional mathematics, it is asserted that such and such exists, in this system it will be asserted that such and such can be constructed. Now it is clear that I will need a more powerful notion of constructibility than that of the Intuitionists if I am to obtain anything like classical mathematics. And since my strategy is to introduce special constructibility quantifiers to be used in my mathematical system in place of the standard existential quantifiers, let us begin by laying out the chief logical features of these quantifiers. This will be done by making use of some familiar semantical notions taken from Kripkean quantified modal logic. It needs to be emphasized however that the semantics of possible worlds, which is to be used here, is brought in merely as a sort of heuristic device and not as the foundations upon which the mathematics of this work are to be based: the constructibility quantifiers are primitives of the system, and the Platonic machinery of Kripkean semantics is used to make the ideas comprehensible to those familiar with this heavily studied area of semantics. The basic idea behind this way of proceeding is this: At one time, Intuitionistic mathematics was completely opaque to the majority of classical mathematicians, because its ideas and procedures were so unorthodox and unfamiliar. However, when classical mathematics was used to analyse and interpret these theories, Intuitionism became much more accessible and interesting to many more mathematicians, even those highly critical of the view.

The constructibility quantifier (Cx) is to be understood as saying, roughly,

It is possible to construct an x such that

but in many contexts, it could also be rendered

It is possible that there be an x such that

depending on what sort of x we are talking about. In the beginning, at least, the second rendition may appear to be somewhat more natural. Thus, in terms of the familiar Kripkean possible worlds semantics,

$-(\mathbf{C}x)(x$ is a tangram in the shape of a circle)

can be understood as saying (but only as a first approximation):

There is no possible world in which a tangram in the shape of a circle exists.

I shall begin with the specification of a formal language, **L**, that has the syntax of a standard first-order language, but whose semantics is quite different. The whole construction will be carried out using such realistic notions as sets, functions, sentence types, etc.

2. The Language **L**

THE SYNTAX OF **L**

The syntax of this language is essentially that of the first-order language of Benson Mates's (El) (except for notational changes and the elimination of sentential letters).

Primitive Symbols

I shall adopt the notational conventions of Mates's book regarding the concatenation of expressions referring to object-language expressions and symbols. Object-language expressions and Arabic numerals will be regarded frequently as autonymous (i.e. as names of themselves). Greek letters will be used as meta-variables.

1. *Variables* Variables are italicized lower-case letters 'u' through 'z', with or without numerical subscripts (the *numerical* subscripts being Arabic numerals of positive integers).

2. *Constants* There are two types of constants: non-logical constants and logical constants. And there are two types of non-logical constants: individual constants and predicates.

    (*a*) *Individual constants* are italicized lower-case letters 'a' through 't', with or without numerical subscripts.

    (*b*) *Predicates* are italicized upper-case letters 'A' through 'Z', with numerical superscripts, and with or without numerical subscripts. (The superscript gives the *degree of the predicate*.)

    (*c*) The *logical constants* are:

$$- \lor \mathbin{\&} \rightarrow \longleftrightarrow (\ ) \; \mathbf{C} \; \mathbf{A}$$

3. *Individual Symbols* An individual symbol is a variable or an individual constant.

Formulas

1. *Atomic Formulas* If φ is a predicate of degree n (or is, as I shall sometimes say, an n-ary predicate), and α_1, α_2, α_3, ... α_n are individual symbols, then

$$\varphi\alpha_1\alpha_2\alpha_3 \ldots \alpha_n$$

is an atomic formula.

2. *Formulas*
   (*a*) All atomic formulas are formulas.
   (*b*) If φ and ψ are formulas and α a variable, then the following are also formulas:

$$-\varphi$$
$$(\varphi\&\psi)$$
$$(\varphi\vee\psi)$$
$$(\varphi\rightarrow\psi)$$
$$(\varphi\longleftrightarrow\psi)$$
$$(C\alpha)\varphi$$
$$(A\alpha)\varphi$$

3. *Further terminology*
   (*a*) I assume the usual definitions of *bound* and *free occurrences of variables* in a formula.
   (*b*) A *sentence* is a formula in which no variable occurs free.

THE SEMANTICS OF L

K-interpretations

A K-interpretation is an ordered quadruple $<\mathbf{W}, \mathbf{a}, \mathbf{U}, \mathbf{I}>$, where:

   $\mathbf{W}$ is a non-empty set;

   $\mathbf{a}$ is a member of $\mathbf{W}$;

   $\mathbf{U}$ is a function which assigns to each member of $\mathbf{W}$ a non-empty set.

I is a function satisfying the following conditions:

(a) it assigns to each individual constant a member of $Z =$ the union of all the $U(w)$, where w is a member of W;

and

(b) if φ is an n-ary predicate, then I assigns to φ, a subset of the n-th Cartesian product of Z with itself.

Comment and terminology. Intuitively, one can think of W as the set of all possible worlds, a as the actual world, and $U(w)$ as the set of things that exist in possible world w. Accordingly, I adopt this way of speaking of the items of a K-interpretation. For individual constant α then, I gives the denotation of α; and where φ is a predicate, I gives the extension of φ. Notice that the denotation of the individual constant α is fixed by I independently of possible worlds. In other words, individual constants function as "rigid designators" (as is customary in Kripkean systems). However, in this system, predicates are also treated as "rigid"—a feature that is not typical. One could obtain a semantical system closer to Kripke's if one provided for the possibility of predicates having different extensions in different possible worlds; but I forgo this complication here, because the theory to be developed later (in Chapter 4) will deal only with "rigid" predicates.

Truth under a K-interpretation

In the following, take M to be a K-interpretation and φ to be a sentence.

(a) If φ is the atomic sentence

$$\theta\alpha_1\alpha_2\alpha_3 \ldots \alpha_n$$

where θ is an n-ary predicate and the α_i are individual constants, then φ is true under M if, and only if,

$$<I(\alpha_1), \ldots I(\alpha_n)> \epsilon I(\theta).$$

(b) If $\varphi = -\psi$, then φ is true under M if, and only if, ψ is not true under M.

(c) If $\varphi = (\psi \vee \chi)$, then φ is true under M if, and only if, ψ is true under M or χ is true under M, or both.

(d) If $\varphi = (\psi \& \chi)$, then φ is true under M if, and only if, ψ is true under M and χ is true under M.

(e) If $\varphi = (\psi \rightarrow \chi)$, then φ is true under M if, and only if, either ψ is not true under M or χ is true under M, or both.

(*f*) If $\varphi = (\psi \longleftrightarrow \chi)$, then φ is true under **M** if, and only if, either ψ and χ are both true under **M** or they are both not true under **M**.

For the quantificational cases, it is convenient to give some preliminary definitions:

(i) *β-variants*

Suppose that $\mathbf{M} = <\mathbf{W}, \mathbf{a}, \mathbf{U}, \mathbf{I}>$ and $\mathbf{M}' = <\mathbf{W}', \mathbf{a}', \mathbf{U}', \mathbf{I}'>$, and that these are both K-interpretations. And suppose that β is an individual constant. Then if **M'** differs from **M** at most in what is assigned to β, then **M'** is a β-variant of **M**.

(ii) If α is a variable, β an individual symbol, and ψ a formula, then $\psi\alpha/\beta$ is the result of replacing all free occurrences of α in ψ by β.

The clauses for the quantifiers can then be given as:

(*g*) If $\varphi = (\mathbf{A}\alpha)\psi$, then φ is true under **M** if, and only if, $\psi\alpha/\beta$ is true under every β-variant of **M**.

(*h*) If $\varphi = (\mathbf{C}\alpha)\psi$, then φ is true under **M** if, and only if, $\psi\alpha/\beta$ is true under at least one β-variant of **M** (where, in the last two clauses, β is the first individual constant not occurring in ψ).

φ is false under **M** if, and only if, φ is not true under **M**.

Comment. Notice that the sentence '$F^1 b$' can be true under a K-interpretation even if $\mathbf{I}(b)$ does not belong to $\mathbf{U}(\mathbf{a})$. Hence, in this system, one can say of an object that does not exist (in the actual world) that it has a property. Similarly, one can say of an object that does not exist in the actual world that it is related to an object that does exist in the actual world.

Terminology

Consequence. A sentence φ is a consequence of a set of sentences Γ if, and only if, there is no K-interpretation under which all the elements of Γ are true and φ is false.

Valid. A sentence φ is valid if, and only if, φ is true under every K-interpretation.

Sound. A system of inference rules is sound if, and only if, any conclusion derived in accordance with the rules is a consequence of the premisses from which it is obtained.

Complete (weak). A system of inference rules is weakly complete if, and only if, every valid sentence can be derived in accordance with the rules from the empty set of premisses.

Complete (strong). A system of inference rules is strongly complete if, and only if, any consequence of any set of sentences can be derived, in accordance with the rules, from the set of sentences.

Rules of Inference

A *derivation* is a finite sequence of consecutive numbered lines, each consisting of a sentence together with a set of numbers (called the *premiss-numbers* of the line), the sequence being constructed according to the following rules:

P (Introduction of premisses). Any sentence may be entered on a line, with the line number taken as the only premiss-number.

T (Tautological inference). Any sentence may be entered on a line if it is a tautological consequence of a set of sentences that appear on earlier lines; as premiss-numbers of the new line take all premiss-numbers of those earlier lines.

C (Conditionalization). The sentence $(\varphi \rightarrow \psi)$ may be entered on a line if ψ appears on an earlier line; as premiss-numbers of the new line take those of the earlier line, with the exception (if desired) of any that is the line number of a line on which φ appears.

E (Existential quantification). The sentence $(C\alpha)\varphi$ may be entered on a line if $-(A\alpha)-\varphi$ appears on an earlier line, or vice versa; as premiss-numbers of the new line take those of the earlier line.

US (Universal specification). The sentence $\varphi\alpha/\beta$ may be entered on a line if $(A\alpha)\varphi$ appears on an earlier line; as premiss-numbers of the new line, take those of the earlier line.

UG (Universal generalization). The sentence $(A\alpha)\varphi$ may be entered on a line if $\varphi\alpha/\beta$ appears on an earlier line, provided that β is an individual constant that occurs neither in φ nor in any premiss of that earlier line; as premiss-numbers of the new line take those of the earlier line.

Terminology

If *n* is a premiss-number of a line, say *m*, in a derivation and φ is on line *n* in that derivation, then φ will be called a premiss of line *m*.

A derivation in which φ is on the last line and all the premisses of that line are members of Γ will be called a derivation of φ from Γ.

Soundness and Completeness of the Inference Rules

That the above system of inference rules is sound can be proved by induction (as in El, ch. 8). That the system is complete can also be easily seen by the aid of some simple lemmas (discussed below).

Notice first that given any sentence φ of **L**, there is a corresponding sentence, consisting of the same primitive symbols in the same order, of a *standard first-order version* of **L**, i.e. a language that has the syntax of **L** but the semantics of standard first-order logic. Let us call the sentence corresponding to φ its *first-order image*. Now, the system of inference rules given above is known to be both sound and complete for the first-order language that is described here. Hence, one can state:

Lemma 1. A sentence φ of **L** is derivable (in **L**) from the empty set of premises if, and only if, its first-order image is derivable (in first-order logic) from the empty set of premises.

Lemma 2. Given any first-order interpretation **N**, there is a K-interpretation $\mathbf{M} = <\mathbf{W}, \mathbf{a}, \mathbf{U}, \mathbf{I}>$ which is such that, for any sentence of **L** φ, φ is true under **M** if, and only if, the first-order image of φ is true under **N**.

Proof. We are going to select a one-world set of possible worlds so that $\mathbf{W} = \{\mathbf{a}\}$, i.e. the only world in **W** is the actual world. We select **U** to assign to the actual world the set = the universe of discourse of **N**. Then let **I** assign to each non-logical constant what the non-logical constant is assigned by **N**. Let **I** assign to each predicate θ what θ is assigned by **N**. It can be straightforwardly verified that for this choice of K-interpretation, what is asserted in the lemma holds.

Weak completeness directly follows. For if a sentence φ is a valid sentence of **L**, then from lemma 2, its first-order image must also be a valid sentence of the standard first-order language. Hence by the completeness of first-order logic, the image is derivable from the empty set of premises. By lemma 1, φ is derivable (in **L**) from the empty set of premises. Strong completeness follows almost as directly. It should be obvious from the above that the set of valid sentences of **L** is not decidable and that compactness holds.

Comment. As I noted earlier, '$F^1 b$' can be true under a K-interpretation even if $\mathbf{I}(b)$ is not a member of the actual world. At first sight, it might appear that one can obtain the absurd result, using Existential Generalization, that if $F^1 b$, then something exists which is F^1. In fact, what one can infer from '$F^1 b$' in this way is '$(\mathbf{C}x)F^1 x$'—which only says that it is possible for there to be an x such that $F^1 x$—something that is not at all absurd. Of course, this brings out the fact that one cannot express ordinary existential quantification in **L**. To do this, we need to turn to a more expressive language.

3. The Language L*

I shall sketch in this section a formal language in which two sorts of quantifiers appear together: the constructibility quantifiers introduced above and also the standard quantifiers of the usual predicate calculus. I do this for purely heuristic reasons—the constructibility theory I shall develop in this work will be specified in another, somewhat more complex, language. The point is: how the constructibility quantifiers interact logically with the standard quantifiers can be more easily grasped within the simpler setting of the language L*.

One can regard L* as dealing with the individual pieces that are used to construct tangrams, on the one hand, and the tangrams that it is possible to construct, on the other. The logic of L* will be two-sorted: there will be two sorts of things talked about in the language; and different kinds of individual symbols will be used to refer to these two sorts of things. Only one sort of thing will be regarded as constructible, so the constructibility quantifiers will be used only to refer to objects of this kind. Furthermore, since we will not be concerned with questions of whether any particular objects of this kind actually exist (i.e. exist in the actual world), only constructibility quantifiers (and no standard quantifiers) will be used in connection with this kind of thing. Of course, when talking about the other kind of thing, standard quantification will be employed. Thus, if one is using L* in the way suggested above, tangrams will be regarded as the constructible sort of thing, and the constructibility quantifiers will be used to assert the constructibility of a tangram of such and such a sort. Standard existential quantifiers will be used to assert the existence (in the actual world) of an individual piece that is used to construct a tangram. In L*, one will assert neither the existence (in the actual world) of any tangram nor the constructibility of any individual tangram piece.

THE SYNTAX OF L*

The primitive symbols of this new language are specified in terms of the primitive symbols of L.

Logical constants: add '∃'

Starred variables are added to the list of variables: take the variables of L and add '*' to get the new variables.

Starred individual constants are added to the list of individual constants. This is done by adding a '*' to each individual constant of **L**.

A *starred individual symbol* is a starred individual constant or a starred variable.

Predicates: For simplicity, I reduce the list of predicates to just one binary predicate 'R^2'.

Formulas

Taking 'variable' to mean 'unstarred variable', 'individual constant' to mean 'unstarred individual constant', and 'individual symbol' to mean 'unstarred individual symbol', the definition of 'formula' is to remain as before with only the following changes: replace the definition of 'atomic formula' with clause (*a*) given below and add to the definition of 'formula' clause (*b*) given below:

(*a*) *Atomic formula*: If α is a starred individual symbol and β is an individual symbol, then

$$R^2\alpha\beta$$

is an atomic formula. Every atomic formula is a formula.

(*b*) If ψ is a formula, α a starred variable, then

$$(\exists\alpha)\psi$$

and

$$(\alpha)\psi$$

are formulas.

THE SEMANTICS OF L*

One should think of there being two distinct universes of discourse for each possible world, one for each type of individual symbol. The starred individual symbols are to be thought of as referring only to the things in the actual world.

K*-interpretations

A K*-interpretation $= <$**W***, **a***, **U***, **I***$>$ differs from a K-interpretation in the following way:

1. **U*** is a binary function that assigns to each ordered pair $<j, k>$, where j is either '0' or '1', and k is a member of **W***, a non-empty

set.[1] $U^*(0, k)$ is the set of things in world k that the starred variables range over, and $U^*(1, k)$ is the set of things in world k that the unstarred variables range over. (As an example, think of the things in $U^*(0, k)$ as the tangram pieces that exist in world k and the things in $U^*(1, k)$ as the tangrams that are constructed in world k.)[2]

2. I^* assigns to each individual constant some element of $Z =$ the union of all the $U^*(1, w)$, where w is any element of W^*; I^* assigns to each starred individual constant some element of $U^*(0, a^*)$; and I^* assigns to 'R^2' a subset of the Cartesian product of $U^*(0, a^*)$ with Z. (To develop the above example, think of 'R^2' as expressing some relation between the tangram pieces that exist in the actual world and the tangrams that are constructed in the various possible worlds.)

Truth under a K*-interpretation is defined in the previous manner for all cases except for the clauses involving the quantifiers. In particular, the atomic sentence

$$R^2a^*b$$

is true under K*-interpretation $< W^*, a^*, U^*, I^* >$ if, and only if, $< I^*(a^*), I^*(b) > \epsilon I^*(R^2)$. For the quantifier cases in general (and in particular for the cases discussed below), we need to add the requirement that *β is an individual constant of the same type as α*, i.e. α is starred if, and only if, β is starred. Except for this change the clauses for the constructibility quantifiers remain the same. The clauses for the standard quantifiers are essentially the same as those for the constructibility, and can be given as follows:

$(\exists \alpha)\varphi$ is true under M^* if, and only if, $\varphi\alpha/\beta$ is true under some β-variant of M^*.

$(\alpha)\varphi$ is true under M^* if, and only if, $\varphi\alpha/\beta$ is true under every β-variant of M^*.

Discussion. Corresponding to each possible world k, there are two types of things: 0-things and 1-things. Hence, we have two sets of

---

[1] One significant difference between the semantics of L^* and standard Kripkean modal semantics is that all the possible worlds of a K*-interpretation must have non-empty domains, whereas the possible worlds of Kripke's semantics may have empty universes (see Kripke, Semantical, pp. 63 and 69). The rationale for this feature of K*-interpretations will be given in n. 2 below.

[2] It should now be clear why, in a K*-interpretation, we only consider possible worlds that have non-empty domains. Since we are only concerned with what tangrams it is possible to construct, we need only consider those possible worlds in which at least one tangram exists (i.e. is constructed).

things from world k: $U^*(0, k)$ and $U^*(1, k)$. However, as far as what is relevant to truth under a K*-interpretation, the 0-things from all possible worlds other than the actual world are not significant: the domain of the standard quantifiers can be regarded as the set of things in the actual world, i.e. $U^*(0, a^*)$; and each starred individual constant can be regarded as naming some actually existing object, i.e. some element of $U^*(0, a^*)$. The unstarred individual constants, on the other hand, may denote 1-things from any possible world k and the constructibility quantifier can be regarded as ranging over the totality of 1-things from all the possible worlds.

Let us use the tangram example to clarify these ideas. There are, in the actual world, many tangram pieces which can be used to construct tangrams. But in some possible world, there are tangrams pieces that do not exist in the actual world. Interpret 'R^2' to be the relation obtaining between a tangram piece x in the actual world and a tangram y constructed in some possible world when, and only when, the piece in question is a part of the tangram, that is, when y is made up of pieces, one of which is x. Notice that the relation 'R^2', so interpreted, is rigid. If some tangram piece x is in this relation to some tangram y, then x will be in this relation to y no matter what possible world y may be in—otherwise, it would not be y.

The reader can determine that, given the above interpretation, the following can be asserted:

$$(x^*)(Cy)R^2x^*y$$
$$-(Ay)(\exists x^*)R^2x^*y.$$

Consider a somewhat different interpretation. We can regard the history of our world as being made up of a totality of temporal slices of the world. Then, corresponding to each moment of time t, let w_t be the world as it is at t. Then, let W^* be the set of all the w_t. If x is the present moment, let a^* be w_x. In other words, the actual world a^* is just the world as it is at the present moment. Let $U^*(0, a^*)$ be the set of people who exist in a^*, i.e. the set of people actually alive at the present moment. For any possible world k, let $U^*(1, k)$ be the set of people who exist in k. Finally, let $I^*(R^2)$ be the relation

is a (biological) descendant of.

Here again, the relation in question can be taken to be rigid. If x is a descendant of y in this framework, then this relationship will hold independently of the worlds in which x and y are to be found.

The reader can verify that, under this K*-interpretation, the following sentences will be true:

$$(x^*)(\mathbf{C}y)R^2x^*y$$

$$-(\mathbf{A}y)(\exists x^*)R^2x^*y$$

and if one believed the Biblical story, one would also believe that

$$(\mathbf{C}y)(x^*)R^2x^*y$$

is true under this interpretation. (The above two interpretations of **L*** presuppose transworld identity. It should be emphasized, however, that these interpretations are being given for illustrative purposes only. Transworld identity will not be presupposed in the constructibility theory to be presented in Chapter 4.)

Inference Rules

The inference rules are obtained by simply adding to the earlier ones another set of quantifier rules (E^*), (US^*), and (UG^*), with the obvious changes. Thus (US^*) would read:

The sentence $\varphi\alpha/\beta$ may be entered on a line if $(\alpha)\varphi$ appears on an earlier line, provided that β is a starred individual constant; as premiss-numbers of the new line take those of the earlier line.

It would be a simple matter to prove that this new system of inference rules is both sound and complete (where 'sound' and 'complete' are defined for the language **L*** in the obvious way). I give below a brief sketch of how completeness can be proved by modifying slightly the Henkin proof of (E1, ch. 8). The proof proceeds by way of the lemma that every set of sentences of **L*** that is *consistent with respect to derivability* (or d-consistent) is satisfiable. The proof of this lemma can proceed as for the case of first-order logic, except for the part of the proof that shows:

[*] Every maximally d-consistent and omega-complete set Δ is satisfiable.

To prove [*], I specify a K*-interpretation M which is such that

[**] For every sentence S, S is true under M if, and only if, S is a member of Δ.

This is done as follows:

W* is a one-element set;

U*(0, **a***) is the set of starred individual constants;
U*(1, **a***) is the set of unstarred individual constants;
For every individual constant α (starred or unstarred), **I***$(\alpha) = \alpha$;
I*(R^2) is the set of all ordered pairs $<\alpha, \beta>$ which is such that the sentence

$$R^2 \alpha \beta$$

is a member of Δ.

One can then prove [**] by induction on the index of S, where the index of a formula of **L*** is the number of occurrences of quantifiers and/or connectives in the formula.

4. Some Objections Considered

We now have the basic logic of the constructibility quantifiers. Some further clarification of the quantifiers can be obtained by analysing an objection to them raised by Dale Gottlieb. In an earlier work (V-C), I made use of constructibility quantifiers, which were taken to be primitives and given an intuitive interpretation. It was argued in that work, on the basis of intuitive reasoning (and without appeal to the above sort of semantical analysis), that these quantifiers behaved logically much like ordinary quantifiers, at least in the contexts in which they were being used in the book. Gottlieb's objection was directed at my use of these quantifiers. For purposes of simplicity of exposition, I shall discuss this objection in connection with a single sentence containing one of these quantifiers. Let us classify any sentence consisting of the words 'There are no more than', followed by an Arabic numeral, which in turn is followed by the word 'star', as an *S*-sentence. For example, the sentence 'There are no more than 1,000,000 stars' is an S-sentence. Now consider the following sentence

[1] (**C**x)(x is an *S*-sentence and x is true).

As I explicated such sentences, one can regard [1] as saying, informally

[2] It is possible to construct an *S*-sentence that is true.

Gottlieb suggests that we can regard these sentences as asserting

[3] $\Diamond$ ($\exists x$)(x is an *S*-sentence that is true).

Now as [1] is intended, it is true just in case there are finitely many stars. But [3] is true even if, in fact, there are infinitely many stars. For no matter how many stars may in fact exist, it is possible that only finitely many stars exist. Strangely, what Gottlieb concludes from this is not that it is improper to replace by '(Cx)' with '$\diamond$($\exists x$)', but rather that the constructibility interpretation of the existential quantifier that I gave does not "assign truth values in an acceptable manner". It should be clear to the reader, from what has been given above, that Gottlieb's interpretation of my constructibility quantifier, using the possibility operator, is a distortion. In terms of the semantical interpretations developed in this chapter, one can say that [3] is true just in case there is a possible world **w** which is such that there is some *S*-sentence in **w** which is true at **w**. But [1] is true just in case there is a possible world **w** which is such that there is some *S*-sentence in **w** which is true at **a** (the actual world). In other words, to determine the truth of [1], one looks to other possible worlds to see what *S*-sentences are to be found there, but to determine the truth of [3], one looks to other possible worlds not only to see what *S*-sentences are there, but also to see what stars exist there.

It should be emphasized again that the above appeal to possible worlds was made to relate the constructibility quantifiers to familiar and heavily studied areas of semantical research. I, personally, do not take possible world semantics to be much more than a useful device to facilitate modal reasoning. Still, I hope to convince most philosophers by means of such analyses that the predicate calculus I shall be using is at least consistent and that it has a kind of coherence and intelligibility that warrants the study of such systems. Furthermore, the discussion should make it clear that my constructibility quantifiers are very different from the quantifiers of Intuitionistic mathematics, according to which ($\exists x$)*Fx* is assertable just in case there is available a "presentation" of an object b and a proof of *F*(b). Thus the assertion (Cx)*Fx* should not be taken as implying that one has (or even that one could in principle devise) an effective procedure for producing an x such that *Fx*.

Up to now, almost all of my discussion of the non-standard quantifiers of this system has been of the constructibility quantifiers. Yet the deductive system I have described above has two kinds of non-standard quantifiers, one corresponding to the existential quantifier (i.e. the constructibility quantifier) and the other corresponding to the universal quantifier. Why have I said so little about the other kind of non-standard quantifier? It is because I see the modal universal quantifier as defined in terms of the constructibility quantifier according to the following:

$$(A\alpha) \ldots \alpha \ldots = -(C\alpha)- \ldots \alpha \ldots$$

For purposes of formal development, however, I have found it simpler to regard the modal universal quantifier as a primitive of the system.

The basic idea of the approach to be taken in this work can now be put simply as follows: existence in mathematics will always be expressed by means of the constructibility quantifiers. More specifically, the system of mathematics I shall be developing in this work will contain two sets of quantifiers: the standard extensional ones and the modal quantifiers described above.[3] What will correspond to the existential theorems of classical mathematics in this system will always be constructibility theorems expressed by means of the constructibility quantifiers.

At this point, the following objection regarding the constructibility quantifiers may occur to the reader: The English phrase 'It is possible to construct', as it occurs in mathematical contexts, seems to have the ontological implications of the phrase 'There exists'. To know that it is possible to construct a house that meets certain specifications does not guarantee that there really is a house that meets these specifications. However, to know that it is possible to construct a number that meets certain specifications may well be enough to guarantee that there exists a number that meets the specification. One might argue that, in the case of mathematical objects, to be constructible is to be. So what do we gain ontologically in substituting constructibility quantifiers for the standard existential quantifiers of mathematics? I shall take up this objection in the next chapter.

---

[3] Students of modal logic will see that these two quantifiers are related to each other in a way that is similar to the way that the actualist and the possibilist existential quantifiers are related to each other. See Forbes (Modality), p. 35 for an explanation of how those two kinds of quantifiers are defined.

3

Constructibility and Open-Sentences

1. The Constructibility of Open-Sentences

WHAT kind of things are to be said to be constructible by the mathematical theorems in my system? In Intuitionistic mathematics, it is proofs or mental constructions that are asserted to be constructed or constructible. In my system, mathematics will be concerned with the constructibility of open-sentences—indeed, tokens (as opposed to types) of open-sentences will be said to be constructible. Thus, in terms of the Kripkean semantics discussed in the previous chapter, to say that it is possible to construct an open-sentence of a certain sort is to say that in some possible world there is a token of the type of open-sentence in question. But what would it be for some token of some open-sentence type *to exist in a possible world*? Here, we can imagine a possible world in which some people, who have an appropriate language, do something that can be described as the production of the token: they may utter something, write something down, or even make some hand signals. Thus, we need not concern ourselves with questions about the ontological status of tokens: in particular, we need not worry ourselves over whether a series of hand signals is or is not an entity, or whether it is a physical object of some sort. Nor should we worry about the propriety of calling the open-sentence traced in the sand by a child "an object". We can regard talk about the existence of such tokens in a possible world as a *façon de parler* in which one is actually talking about what people do in that world. To say that someone has constructed an open-sentence is not to say that an entity of a certain sort has been constructed but only that the person has done something—he has performed the appropriate series of actions. This is one reason why, when the objects we are discussing are open-sentence tokens, I prefer the 'It is possible to construct' reading to the 'It is possible for there to be' reading. Still, it is useful to treat open-sentence tokens as ordinary objects, especially when one makes use of the possible worlds semantics presented earlier.

Let us reconsider the objection with which the previous chapter ended. It should now be clear that the sorts of things which my system asserts to be constructible are much more like tangrams and houses than sets and numbers. To say that it is possible to construct an open-sentence token of such-and-such a kind is not to say that anyone has actually constructed such an open-sentence token, nor does it say that such-and-such a token actually exists. Indeed, in the sense in which I use the constructibility quantifiers, I do not know what it would mean to assert that it is possible to construct a number or a set.

2. A Simple Type Theory for Open-Sentences

I wish now to characterize the kinds of open-sentences to be discussed in the theory to be developed. As my theory can be regarded as a kind of simple type theory, this can best be done by contrasting my approach with a more traditional theory of types. Let us consider first of all Frege's system, in which there are basically two sorts of "things" treated: *objects* and *concepts*.[1] Concepts are not open-sentences, but they are related to open-sentences much in the way objects are related to names, if one thinks of open-sentences and names as being linguistic items and concepts and objects as being "in the world". Concepts are sorted by Frege into levels and kinds; but for purposes of simplicity and perspicuity, let us restrict our consideration for now to just the monadic concepts that objects are said to "fall under". Now the sentence 'François Mitterrand is president of France' can be analysed as consisting of two parts: a name, 'François Mitterrand', and an open-sentence, 'x is president of France'. The name has a reference: it denotes an *object*, namely the person who is so named. The open sentence, according to Frege, can also be regarded as having a reference; it too denotes something, namely a *concept*. From the point of view of logic, the principal relation that obtains between objects and concepts is the "falling under" relation. Thus, when the variable 'x', occurring in the above open-sentence is replaced by the name 'François Mitterrand', we obtain a true sentence. In that case,

---

[1] See Frege's (Laws), sects. 1–4 for his explanations of the basic notions of his system. For the most part, I follow the interpretation given by M. Furth in his introduction to Frege's (Laws). See also my (V-C), ch. 1, sect. 7, and my (Ramsey) for more details on how I interpret Frege's system.

we can say that the object Mitterrand satisfies the open-sentence and also falls under the concept denoted by that open-sentence. Thus, the monadic concepts described above correspond to open-sentences with only one free variable ranging over the totality of objects; and the relation of *falling under* corresponds to the relation of satisfaction that things have to open-sentences; so if an object falls under a concept, we can regard the concept as a sort of property that the object has. Henceforth, if an object x falls under concept F, x will be said to have the property F; if x does not fall under F, then x will be said to lack the property F. For example, consider the concept *being made of wood* that corresponds to the open-sentence 'x is made of wood'. The chair in my office falls under the concept in question and hence will be said to have the property of being made of wood. The above open-sentence, on the other hand, will be said to be true of (or satisfied by) the object in question.

In Frege's system, there is a third sort of logical entity talked about, namely extensions of concepts. Extensions were classified by Frege as a kind of object, but they are certainly sufficiently special, from my point of view, to be distinguished from the ordinary physical objects that one usually thinks of when asked to give an example of an object. The extension of a concept can be regarded, for most purposes, as the class of things that fall under the concept. Hence, the chair in my office is related to the three different kinds of logical items in Frege's system as follows:

　(*a*)　it falls under the concept of being made of wood;
　(*b*)　it satisfies the open-sentence 'x is made of wood';
　(*c*)　it is a member of the extension of the concept of being made of wood.

Here, the reader may wish to pause to ponder the question of just why it is necessary to postulate all three sorts of logical items. Might not one get by in logic and mathematics with only two, or possibly even only one of these items? In fact, might not just open-sentences do the work of all three? If not, why not?

Now the structure of Frege's system of monadic concepts is even more complex than I have indicated thus far, since it is not only objects that can fall under concepts. The type of concept I have been discussing thus far can also fall under higher-level (second-level) concepts. For example, the concept of being made of wood is a first-level concept that falls under the second-level concept of *having at least one object falling under it*. Second-level concepts in turn may fall under various third-level concepts. For example, the above-mentioned second-level concept falls under the third-level concept of *having at least one first-level concept falling under it*. It is easy to see that one can continue in this way to obtain concepts of any finite level. The levels being discussed here are regarded by Frege as being exclusive, i.e. no monadic concept belongs to more than one level. Using the property terminology, the following structure emerges: we have a hierarchy in which objects are to be found at the lowest level; at level 1, there are properties of objects; at level 2, there are properties of properties of objects; etc. In short, we have the structure of Simple Type Theory. Indeed, if we replace the *falling under* relation in this structure with the *membership* relation, and change properties to sets, so that, for example, a property of objects is replaced by the set of those objects that have the property, what we get is the familiar set-theoretical structure of Simple Type Theory. (See my (Ramsey) for more details.)

We have been considering here only monadic concepts. Frege's hierarchy of concepts can be seen to be considerably more complex when relations (i.e. n-adic concepts) are brought into consideration. However, these complications need not detain us here since they introduce no fundamentally new ideas and since, in the system to be developed in this work, the usual theory of relations will be constructed in terms of properties, in a sense providing a sort of ontological reduction of relations to properties.

As is well known, Frege's system was seriously flawed: the set of axioms upon which Frege hoped to establish the fundamental laws of arithmetic was shown by Bertrand Russell to be inconsistent.[2] This gave rise to the problem of finding a remedy: How can Frege's logical system be made consistent, while at the same time preserving the essential features of his development of arithmetic from the axioms? Recall that there were three different kinds of logical items in Frege's system that were closely related: open-sentences, concepts, and extensions of concepts. Extensions of concepts were postulated, it would seem, primarily in order to have something to serve as the referents of numerical terms. What then are numbers? As Frege analysed sentences containing numerical terms, numbers had to be

---

[2] Frege (Laws, App. II) contains Frege's reaction to Russell's paradox: "Hardly anythi ɔg more unwelcome can befall a scientific writer than that one of the foundations of his edifice be shaken after the work is finished". It also gives Frege's unsuccessful attempt to repair the system.

objects—not concepts. So he needed some sort of object to serve as the referents of these numerical terms. Thus, although he could define the second-level cardinality properties expressed by:

F has cardinality 0

F has cardinality 1

F has cardinality 2

.

.

.

without reference to extensions of concepts, Frege felt the need to have the terms '0', '1', '2', ... function as names of objects. So he selected the extensions of these second-level properties to be the natural numbers. For example, in Frege's system, the number one is just the extension of the property of having cardinality one. But, as Russell showed with his "no-class" theory, one can get along in mathematics without postulating any extensions or classes.[3] In Russell's system, in place of statements about natural numbers, one has statements about the corresponding cardinality properties. This suggests a simple way of obtaining from Frege's system a consistent system of types which can be used to develop mathematics along the lines Frege had mapped out. Basically, the idea is to eliminate from the deductive part of Frege's system all those axioms that involve the notion of extension, while making use of Frege's logical hierarchy of concepts and almost all of his logical laws. This would involve discarding the infamous Axiom V, which plays a crucial role in the derivation of Russell's Paradox, and it would involve dropping from the ontology of Frege's system the third category of logical items described above, namely extensions of concepts. What would remain would be a deductive system powerful enough to yield the simple type-theoretical version of classical mathematics, provided that we add an Axiom of Infinity to the set of axioms.[4] Such a system can accommodate the essentials of the Logicist's analysis of classical mathematical reasoning and applications in science. When the formal system is interpreted classically (so as to reflect the intended Fregean

---

[3] See my (V-C), ch. 1, sect. 4.

[4] See my (Ramsey) for a fuller discussion of this fact.

ontology), the predicate variables take as values Fregean concepts, and the existential quantifier is explained in the standard referential way. So for each type of concept, we have standard existential quantification over the totality of concepts of that type. All of these existential quantifiers will be replaced in my system by constructibility quantifiers involving variables for open-sentences of the corresponding type. Essentially, then, a phrase of the form 'There is a concept F such that' in the classical system gets replaced in my system with the corresponding phrase of the form 'It is possible to construct an open-sentence F such that'. The kinds of open-sentences I shall be talking about in my system will be classified into types along the lines of the Fregean system. Thus, for each predicate variable of the formal system, there corresponds, under the classical system, a type of concept, whereas under my interpretation, there corresponds a type of open-sentence that is constructible. For example, corresponding to the assertion in the classical system that there is a monadic concept of level 1 which all objects fall under, there is the assertion in my system that it is possible to construct a monadic open-sentence of level 1 which all objects satisfy. More generally, where the classical system asserts the existence of a level-1 monadic concept satisfying some condition, my system will assert the constructibility of a level-1 monadic open-sentence satisfying the analogous condition. In this system, then, only the open-sentence survives from Frege's trio of logical items to which ordinary objects were supposed to be related. And one never asserts the existence of open-sentences in this system: one only asserts the possibility of constructing them.

3. A Comparison with a Predicative System

The previous discussion suggests that on my view, for every Fregean first-level monadic concept F, it is possible to construct an open-sentence which is satisfied by just those objects that fall under F. This suggests that it is possible to construct at least as many first-level open-sentences as there are first-level monadic concepts. But are there not uncountably many such concepts? How can it be possible to construct uncountably many open-sentences? And aren't we going to run into semantical paradoxes due to *impredicativity* and self-reference in trying to define the satisfaction conditions of all these open-sentences? These questions can be clarified by relating the present approach to constructibility with the approach taken in my earlier book, (V-C), in which I presented an interpretation of a predicative system of set theory using constructibility quantifiers. The earlier work differs from the present one in that, in the earlier work, each constructibility quantifier was associated with an effective rule for generating the open-sentences in question. One could construe the quantifiers as saying, in effect, 'It is possible to construct an open-sentence following such-and-such a procedure.' But more importantly, all the open-sentences talked about were open-sentences of, essentially, the language of the set theory itself and the satisfaction conditions of the open-sentences treated were given in terms of features of the system—features that were linked to the predicative nature of the set theory. Naturally, only countably many of these open-sentences could be constructible. But the present approach is not so constrained. Thus, the earlier approach required the assertion 'It is possible to construct an open-sentence satisfying condition C' to be true just in case (to revert to the K-interpretation method of analysing the constructibility quantifiers) there is a possible world in which someone constructs an open-sentence which not only satisfies C, but also is essentially an open-sentence of the particular language presented in the book. The present approach places no restriction on the language to be employed in constructing the open-sentence—in particular, there is no requirement that we understand the language or that we be able to grasp or to define the satisfaction condition of the open-sentence constructed. Thus, we obtain a more powerful system of constructibility that carries us beyond the predicativity of the earlier system.

It might then be asked how I can speak of satisfaction in the case of open-sentences from some completely unknown language. How is satisfaction to be defined? Obviously, I cannot define satisfaction for the open-sentences of a completely unknown language. Instead, satisfaction is taken to be a primitive notion of our overall theory, much as membership is in systems of axiomatic set theory. It should be emphasized, however, that a wide variety of things are known about satisfaction, in virtue of which the notion is linked conceptually with other items in the theory. We have knowledge of a specific nature regarding satisfaction. For example, we know such things as that this pen is related by the satisfaction relation to the open-sentence 'x is a writing instrument.' We also have knowledge of more general matters involving satisfaction, as for example that, for every open-sentence F

of level 1, it is possible to construct an open-sentence of level 2 that F satisfies and that no open-sentence not extensionally identical to F satisfies.

Readers of my previous book may notice that I do not call my present theory 'nominalistic'. My reasons for eschewing this label are connected with the fact that there are two key ideas underlying nominalism. One is ideological; the other is ontological. From the ideological point of view, only certain *notions* of mathematics are permitted by the nominalist. For example, the membership relation of classical set theory is regarded as illegitimate. From the ontological point of view, only certain sorts of *objects* are permitted into the ontology of the nominalist's theories. Historically speaking, those who have been ontological nominalists have also been ideological nominalists, but one can be an ontological nominalist without being an ideological one. Thus, my constructibility system is, from the ideological point of view, not nominalistic—after all, the idea of satisfaction being presupposed by the system is a very strong one, being little different in strength from the classical set theorist's idea of membership. Still abstract objects are not asserted to exist in my system. Those who believe that an ontological nominalist must be an ideological nominalist evidently come to this belief because they think that one cannot make sense of certain sorts of ideas (such as the idea of satisfaction mentioned above) unless one accepts a Platonic ontology. But this, I have been arguing, is a mistake.

4. The Logical Space of Open-Sentences

The constructibility quantifiers are to be understood as saying 'It is possible to construct an open-sentence such that . . .'. This may lead some readers to suppose that an assertion of constructibility carries with it the implication that the asserter knows how to construct such an open-sentence. This, as I have taken pains to point out, would be a mistake. A related mistake is to assume that the use of these quantifiers requires that one be able to determine if a given open-sentence token is satisfied by some object, if some series of marks is or is not an open-sentence token, or even how to interpret a sequence of signs that is in fact an open-sentence token. It needs to be emphasized that the constructibility theory does not concern itself with many aspects of open-sentences and the things that satisfy them. Consider a theory of constructibility that is, in certain respects, quite similar: Euclid's geometry. Here, we have a well-developed mathematical theory that is concerned with constructing lines and points in space. It tells us, for example, that given any straight line segment, it is possible to construct a perpendicular bisector of this line segment. But notice that it does not tell us how to construct straight lines or even how to determine if a given line is or is not straight.

One way of seeing more clearly how I view the constructibility of open-sentences is to see how the traditional existential way of interpreting Simple Type Theory can be regarded from the point of view of the present work. Take the case in which the quantifiers are treated in the standard extensional way and the predicate variables are said to range over sets. Now can we not regard each of these sets as a sort of "place" for an extension of a possible open-sentence to fit into—a "place", so to speak, in the "logical space of open-sentences". Putnam made a closely related point when he wrote:

[W]e *can* take the existence of sets as basic and treat possibility as a derived notion. What is overlooked is that we can perfectly well go in the reverse direction: we can treat the notion of possibility as basic and the notion of set existence as the derived one. Sets, to parody John Stuart Mill, are permanent possibilities of selection (What, p. 71).

Thus, in terms of this idea, when we say it is possible to construct an open-sentence with such and such an extension, we can be regarded as saying that there is room in the logical space of open-sentences for such an open-sentence. Similarly, to assert the constructibility (in my sense) of a tangram of such-and-such a sort is not to assert that one knows how to construct such a tangram or that one has a method for constructing such a thing: the constructibility in question depends only upon there being room in space for a tangram of that kind to be constructed. Impredicativity is not a problem here since we are not attempting to "construct" the open-sentence in question: we are not attempting to define it; nor are we trying to specify their satisfaction conditions. We are not even saying that the conceptual resources of some theory (explicitly or implicitly given) are sufficient to define the satisfaction conditions of such an open-sentence. Thus, the present theory is very different from the constructibility theory of (V-C) in so far as no attempt is made in the present theory to define the satisfaction conditions of the open-sentences which are said to be constructible.

Further light can be shed on the present theory by comparing it to Euclid's version of geometry, which, as I mentioned earlier, is also concerned with the constructible.[5] Euclid's theory can be easily converted into something very close to present-day versions of Euclidean geometry by simply replacing the constructibility of lines and points by the existence of such things. Indeed, we can move from constructibility to existence in easy steps. To say that it is possible to construct a line satisfying such-and-such conditions is very much like saying that such a construction is not precluded by the nature of the *geometrical space* one is discussing—in other words, that there is room in the geometrical space for such a line—or in other words yet, that there exists a place for the line. Then, it is a small step from the existence of a place for a line to the existence of the line itself. But if we can move from constructibility to existence in such easy steps, it is not surprising that we can reverse the procedure and move from existence to constructibility, as we shall see shortly in the case of Simple Type Theory.

I hope it is clear from my overall position that, in speaking as I do above of the logical space of open-sentences, I am not postulating some new sort of mysterious entity. Such talk is meant to function much as does our talk of possible worlds when discussing modality: I see the notion of a logical space of open-sentences as a sort of heuristic device only. The theory of constructibility that emerges, however, is meant to provide the basis for representing, analysing, and understanding many aspects of the world, as well as our reasoning about the world.

5. The Kinds of Open-Sentences to be Discussed

Viewing the system under consideration as a sort of theory about the "logical space of open-sentences" makes it easier to understand how I can be choosy about the kind of open-sentence my theory will treat.

---

[5] As J. J. M. Bos describes Euclid's *Elements*, "constructions with straight lines and circles were legitimized by Postulates 1–3" (Equations, p. 336). S. Bochner, commenting on the omission of the fundamental theorem of prime numbers from Euclid's treatise on the theory of numbers, writes: "The Greeks never quite arrived at the insight that in mathematics it might be *conceivable* to have an 'existency' which is *totally* divorced from 'constructibility'; and the fundamental theorem is an existence theorem not from choice but from necessity (Role, p. 215).

It is useful to restrict one's consideration to only those sorts of open-sentences that will reveal the essential features of this "logical space" and to prune away those sorts of open-sentences that make forming generalizations complicated or messy. Indeed, I will progressively delimit the sorts of open-sentences that the theory will consider. Thus, I shall not feel compelled to deal with ill-behaved or paradoxical open-sentences in this system. I start right off by stipulating that by 'first-level open-sentences' I mean only those open-sentences F that are "well-defined" over the range of objects, i.e. those level-one open-sentences that are such that, for each object x, either x satisfies F or x does not satisfy F. Similarly, for level-2 open-sentences: we want only those level-2 open-sentences F that are such that for each well-defined open-sentence x of level 1, x either satisfies F or x does not satisfy F, and we want this to be so independently of what possible world x may be in. The same goes for level-3 open-sentences, and so on. In other words, the constructibility quantifier '(Cx) $(\ldots x \ldots)$' is to be understood as saying that it is possible to construct an open-sentence x that is "well-defined" over the previous-level open-sentences and that is such that $\ldots x \ldots$.

Consider the following example of a paradoxical open-sentence, taking the token written on this very page as the open-sentence about which we raise questions,

　　　x does not satisfy itself

and ask if it does or does not satisfy itself. If it does, it does not; and if it does not, it does. Or at least, so it would seem. But there is no good reason why the constructibility theory should be required to deal with such open-sentences. The constructibility theory is concerned with the logical space of well-defined open-sentences only.

Imagine, this time, a token of the open-sentence

　　　x weighs more than Chihara does

constructed in a possible world in which I weigh only 99 pounds. Does the one-hundred-pound sack of sugar in the grocery store across the street in the actual world satisfy that token open-sentence? Which Chihara is being referred to in that open-sentence? The one in the actual world or the one in that possible world? Does it matter as far as the constructibility theory goes? Not at all.

6. Quine's Objections to Modality

Since Quine rejects the use of modalities in scientific theories, he would clearly reject my use of the constructibility quantifiers. Quine's principal objections to the use of modalities are based on the problems of referential opacity. Thus, Quine has written,

Kaplan wonders at the asymmetry between my attitude toward belief and my attitude toward modal logic . . . I had a reason. . . . It was that the notion of belief, for all its obscurity, is more useful than the notion of necessity. For this reason my treatment of modal logic was brief and negative; I was content to outline the opacity problems (Kaplan, pp. 343–4).

However, I side with a growing number of philosophers who do not regard the puzzles about referential opacity as a sufficient reason for eschewing all use of modalities in science or philosophy. First of all, practically every fundamental notion used in philosophy and science gives rise to deep conceptual problems. Such notions as truth, satisfiability, evidence, confirmation, verification, belief, utility, and probability all have conceptual problems connected with them. I take it that no philosopher seriously recommends eschewing all uses of, say, probability in science. Besides, even if the general notion of possibility gives rise to problems of referential opacity, when the notion of possibility is restricted to just the constructibility quantifiers, as it is in my system, the formal logic of the system is not undermined by opacity problems and can be clearly set out with the standard sort of rules of derivation. Indeed, as will become evident in the next chapter, the constructibility system to be developed there will not be troubled by the notorious problems of "transworld identity".[6]

Ernest Adams once wrote in (Topology) regarding Euclid's constructibility version of geometry: "Euclidean geometry cannot be criticized for lack of rigor simply on the grounds of its modal formulation, whatever other faults it may have in this respect, since the logical laws to which the constructibility quantifier conforms are quite clear" (Topology, p. 406). One might say the same of the constructibility quantifiers of the present system.

As for Quine's suggestion (in the above quotation) that modal

---

[6] For a discussion of these problems, see G. Forbes (Modality), ch. 3, and Plantinga (Transworld). Quine's objections to modal logic, based on the opacity problems, also involve raising problems of transworld identity. See Paul Gochet (Ascent), ch. 7, for a recent discussion of these objections.

notions are not useful, the following quotation from Putnam provides, I believe, a more accurate appraisal of the usefulness of modality in science:

> From classical mechanics through quantum mechanics and general relativity theory, what the physicist does is to provide mathematical devices for representing all the *possible*—not just the physically possible, but the mathematical possible—configurations of a system. Many of the physicist's methods (variational methods, Lagrangian formulations of physics) depend on describing the actual path of a system as that path of all the *possible* ones for which a certain quantity is a minimum or maximum. Equilibrium methods in economics use the same approach. It seems to us that 'possible' has long been a theoretical notion of full legitimacy in the most successful branches of science. To mimic Zermelo's argument for the axiom of choice, we may argue that the notion of possibility is intuitively evident and necessary for science (What, p. 71).

Putnam goes on to say: "It seems to us that those philosophers who object to the notion of possibility may, in some cases at least, simply be ill-acquainted with physical theory, and not appreciate the extent to which an apparatus has been developed for *describing* 'possible worlds'" (p. 71).

It also may be mentioned, in defence of using the constructibility quantifiers in mathematics, that geometry was practised for over two thousand years within the framework of Euclid's system[7]—a system that, as I emphasized earlier (in Section 4), is constructive in nature. I believe that it is fortunate for mathematics that no influential philosopher had convinced Greek mathematicians to eschew all theories tainted with modality.

There is another objection Quine raises to the use of modalities. He sometimes argues that one must formulate one's theories in the standard extensional language of first-order logic if one is to disclose the ontic commitments of one's views. And he has suggested that to refuse to formulate one's scientific and philosophical views in such an austere language is, in effect, to turn one's back on ontology altogether. (See, for example, his (Facts), p. 161.) There seem to be

---

[7] By the late sixteenth century, some mathematicians were willing to supplement Euclid's postulates with new ones that would explicitly permit geometrical constructions going beyond the ruler and compass constructions that were justified by the first three postulates. F. Viète, for example, advocated adding a postulate that would allow "neusis constructions" (Bos (Equations), p. 336). The new postulate, however, was still a constructibility postulate.

two basic ideas behind this doctrine: (1) By expressing one's views in the language of first-order logic, one makes explicit one's ontic commitments, in so far as it can be determined, with relative ease. What sorts of things are being presupposed by these views? One need only determine what sorts of things would have to belong to the range of the bound variables in order that the assertions of the view be true (see my (V-C), ch. 3, for details). (2) When one's views and beliefs are expressed in natural languages or in languages with modalities, the simple test of ontic commitment can no longer be applied, and the question as to whether or not the views or beliefs presuppose entities of such-and-such a kind becomes impossible or meaningless.

I will allow that it is generally easier to determine what sorts of things are being presupposed by a theory when it is formulated in the language of the predicate calculus than when the theory is given in a natural language. But the question of whether some abstract theory presupposes or implies the existence of mathematical objects is not what is at issue here. For I am concerned with the question of whether real human beings, living today and with the science we have today, ought to believe in the existence of mathematical objects. Quine suggests that we can answer this latter question by imagining that all of a person's beliefs about traits of reality worthy of the name be expressed in his austere canonical notation and by applying his test of ontic commitment to this imagined theory. (See Quine's (W and O), pp. 221, 228.) But why should we suppose that all of a person's beliefs about traits of reality worthy of the name *can be formulated in such a language*? Furthermore, even if one succeeded in constructing a scientific theory in Quine's canonical language, how could one be sure that *all of one's beliefs* about reality had been captured in this theory?

Well why is it important that all of the person's beliefs about reality be expressed by the theory in question? Because if only some of the person's beliefs about the world are formulated in the canonical language, then it becomes doubtful that we could use Quine's simple test of ontic commitment in the way he suggests in his writings in the philosophy of mathematics. Thus, suppose scientist X formulates a theory T in Quine's canonical notation. And suppose that X believes T to be true, and that by Quine's criterion, T does not presuppose the existence of, say, molecules. Can we conclude that X's views about the world do not presuppose the existence of molecules? Of course not. For X may have other beliefs that, in conjunction with T or some other theory that X accepts, imply or presuppose the existence of molecules. Thus, in order to determine with confidence, in the way Quine suggests, what sort of entities a person's views and beliefs presuppose, we need to be assured that all of the person's views and beliefs about reality get set down in this canonical notation. But why should we suppose that all our views about the world can, in fact, be formulated in Quine's canonical notation? And how can one tell that all of one's beliefs about reality have been set down in any such theory? The vaunted clarity of Quine's method seems to me to be largely illusory. And besides, forcing a person to formulate her beliefs within the restricting confines of first-order logic may produce all sorts of distortions and inaccuracies in our perception of what the person actually believes about reality.

As for the second point, I do not agree that one can only make a rational assessment of what sorts of entities are being presupposed by a person's views about the world by getting the person to formulate her views in Quine's canonical language. In philosophy, as in life, we should do the best we can in the actual situation in which we find ourselves. Recall that I am concerned with the question of whether we ought to believe in the existence of mathematical objects. Ontological questions of this sort have been discussed in the past, with great success, without resorting to the Quinian criterion of ontic commitment. For example, physicists and chemists debated the question of whether it is reasonable to believe in the existence of molecules, without resorting to reformulating their scientific theories in Quine's canonical notation (for details, see Mary Jo Nye's (Molecular)). I see no reason why we cannot do so in the case of mathematical objects.

4

The Deductive System

I CAN now give the principal features of the deductive system within which my constructibility theory of open-sentences is to be developed. I begin with the overall logical framework. An omega-sorted first-order language will be specified.[1] Accordingly, there will be infinitely many different types of individual symbols corresponding to the infinitely many levels of Frege's system. The language will be given a possible worlds semantics of the sort discussed in Chapter 2.

1. The Language **Lt**

This logical language will be specified by giving the respects in which it differs from **L***.

THE SYNTAX OF **Lt**

Variables. These are the italicized lower-case letters '*u*' through '*z*', with or without numerical superscripts, and with or without numerical subscripts.

Individual constants. These are the italicized lower-case letters '*a*' through '*t*', with or without numerical superscripts, and with or without numerical subscripts.

The superscripts are used to indicate the *level* of the individual symbol. Thus, a variable with superscript 5 will be called 'a level-5 variable'. An individual symbol without any superscripts will be regarded as a level-0 individual symbol.

Predicates. For the sake of simplicity of exposition, I have reduced the

---

[1] The formalized theory set forth in this chapter differs from the one I sketched in (Simple) in several respects: first, the language of the present system is a first-order omega-sorted language, as opposed to the omega-level predicate calculus of the earlier system; and second, the open-sentences talked about in the present system are all monadic, whereas the theory described in (Simple) was concerned with *n*-adic open-sentences.

predicates to be included in the vocabulary to just two kinds of binary predicates: S-predicates and I-predicates. (In a fuller exposition, I would include an infinite number of predicates.) Later on (when I present the theory **Ct**), I will interpret the S-predicates to be satisfaction and all but one of the I-predicates to be extensional identity (the one exception standing instead for strict identity). The predicates, then, are italicized upper-case letters '*S*' and '*I*', with or without numerical superscripts. Since all predicates in the language are to function as binary predicates, there is no need to use superscripts to indicate the degree of the predicate. Instead, I shall use superscripts to give the *level of the predicate* (no superscript indicating level 0).

Atomic formula. If θ is the S-predicate of level n, α is an individual symbol of level n, and β is an individual symbol of level $(n+1)$, then

$$\theta\alpha\beta$$

is an atomic formula.

If θ is the I-predicate of level n, and both α and β are individual symbols of level n, then

$$\theta\alpha\beta$$

is an atomic formula.

Formula. The definition of 'formula' is as before, except for the clauses involving quantification.

If φ is a formula and α is a variable of level n $(n>0)$, then

$$(\mathbf{C}\alpha)\varphi$$

and

$$(\mathbf{A}\alpha)\varphi$$

are formulas.

If φ is a formula and α is a variable of level 0, then

$$(\exists\alpha)\varphi$$

and

$$(\alpha)\varphi$$

are formulas.

THE SEMANTICS OF Lt

A *Kt-interpretation* is an ordered quadruple $<W, a, U, I>$ in which:

W is a non-empty set—the set of possible worlds;

a is a member of **W**—the actual world;

U is a function that assigns a non-empty set to each ordered pair $<n, w>$, where n is an Arabic numeral of a non-negative integer, and **w** is a member of **W**; U(n, **w**) can be regarded as the set of things of level n that exist in **w**;

I is a function that assigns:

    (*a*) to each individual constant of level 0, some element of U(0, **a**);

    (*b*) to the S-predicate of level 0, some subset of the Cartesian product of U(0, **a**) with $Z[1] =$ the union of all the U(1, **w**), where **w** ϵ **W**;

    (*c*) to the I-predicate of level 0, some subset of the Cartesian product of U(0, **a**) with itself;

    (*d*) to each individual constant of level n $(n > 0)$, some element of $Z[n]$;

    (*e*) to the S-predicate of level n $(n > 0)$, some subset of the Cartesian product $Z[n] \times Z[n+1]$;

    (*f*) to the I-predicate of level n $(n > 0)$, some subset of $Z[n] \times Z[n]$.

Comment. As for the case of the language **L**, the predicates in this language are taken to be "rigid", so that we do not have to consider different extensions of these predicates in different possible worlds. Why I can make use of this simplification of the semantics of this language will be explained shortly.

Rules of Inference

The inference rules are as before, except for minor changes in the quantifier rules (E), (US), and (UG). For example, (US) would now go:

The sentence $\varphi\alpha/\beta$ may be entered on a line if $(A\alpha)\varphi$ appears on an earlier line, provided that β is an individual constant of the same level as that of α; as premiss numbers of the new line take those of the earlier line.

The sentence $\varphi\alpha/\beta$ may be entered on a line if $(\alpha)\varphi$ appears on an earlier line, provided that β is an individual constant of level 0; as premiss numbers of the new line take those of the earlier line.

It can be seen that these rules provide us with a sound and complete system of derivations. However, these rules clearly need to be augmented to make the system usable as a basis for the development of mathematics: we need to add axioms. In short, we need a theory.

2. The Theory **Ct**

The Constructibility Theory of Types (or **Ct** for short) will be a deductive theory formalized in the language **Lt**. Its vocabulary will consist of all the S-predicates and I-predicates. Hence, a formula of **Ct** will be a formula of **Lt** in which the only non-logical constants occurring in it are S-predicates and/or I-predicates. In what follows, whenever I speak of formulas or sentences, I shall mean formulas or sentences of **Ct**, unless it is explicitly said otherwise. It should be noted that individual constants are not part of the vocabulary of **Ct**. Thus, no theorem of the theory will contain an occurrence of any individual constant. However, individual constants are used (as parameters) in formal derivations of theorems of **Ct**.

The non-logical constants of this theory are to be interpreted as follows: (1) All the S-predicates stand for *satisfaction*. Thus,

$$S^1 a^1 b^2$$

is to be understood as saying that a^1 satisfies b^2. (2) The I-predicate of level 0 stands for *strict identity*; all the other I-predicates stand for *extensional identity*, i.e. an open-sentence F is extensionally identical to open-sentence G if, and only if, F is satisfied by exactly the same things (in the range of the argument variable) that G is.

Rigidity

Are the relations, as they are interpreted above, "rigid"? Consider, first, the I-relation of level 0. This relation (strict identity) is confined to objects in the actual world; so there is no need to consider its extension in different possible worlds. Clearly, this relation is rigid. What about the I-relations of higher levels? In this case, we are concerned with extensional identity, which is defined in terms of satisfaction. But, as I shall argue below, satisfaction, as the term is used in this system, is rigid; so extensional identity is too.

Turning now to the satisfaction relations, recall, from the previous chapter, that this theory is concerned only with open-sentences that are "well-defined" over the appropriate argument ranges; so any possible open-sentence token F that is of concern in this theory must be such that each thing in the appropriate argument range of F either does or does not satisfy F. We thus avoid any problems that might arise if we allowed there to be, in some world w_1, an open-sentence that satisfied a higher-level open-sentence X, but that, in another world w_2, did not satisfy X. Such an X would not be well-defined over its argument range. Hence, we have reason for taking satisfaction to be rigid.

However, further investigation suggests that there may be a problem with this conclusion of rigidity. Suppose, as may well be the case, that no one has ever constructed an open-sentence that is true of just the following three items: Coit Tower, my left thumb, and my cat. Still, it is possible to construct such an open-sentence. How should this possibility be described in terms of the possible worlds model? Clearly as follows: There is a possible world w in which someone constructs an open-sentence token that is true of the three items mentioned. Thus, the satisfaction relation with which we are concerned here holds between the open-sentence token in w and the three objects in the actual world. Basically then, when we are concerned with the possibility of constructing a level-one open-sentence that is true of objects in the actual world, we look into the various possible worlds to see what open-sentences are to be found there; but, in so far as satisfaction is the crucial relationship here, we treat these open-sentences as if they occurred in the actual world, ignoring what objects in the other possible worlds may satisfy these open-sentences. Thus, what is relevant to our concerns is not whether the open-sentence is true of various objects in the world in which the open-sentence is to be found, but only whether or not the open-sentence is true of various objects in the actual world.

Consider now the following problem: Imagine that there is some open-sentence token that appears in possible worlds w_1 and w_2, but that the one in w_1 is true of all three of the mentioned items, whereas the one in w_2 is true of only my cat. Will this not require us to regard satisfaction at least as a function of the various possible worlds in which the open-sentence occurs? One natural way of responding is to say that the open-sentence in question is not "well-defined". But such a response may not be entirely convincing. For, given any open-sentence token S that is true of the three items mentioned above, it seems conceivable that this very token could appear in another world with a sufficiently different background of linguistic practices, so that the token is satisfied by only my cat.

Remember, that for me, this whole possible world structure is an elaborate myth, useful for clarifying and explaining modal notions, but a myth just the same. It would be a mistake to take this myth too seriously and imagine that we are exploring real worlds, finding there real open-sentence tokens that have puzzling features. Look at it this way: I am constructing a kind of model. And as a model constructor, I can make certain stipulations to facilitate using the model. Thus, why not take the difference in extension described above as sufficient to conclude that it is not S that appears in the other world? After all, open-sentence tokens are "logical constructions". Someone makes certain sounds, draws in the sand, or moves her hands and fingers in certain ways, and we say that an open-sentence has been produced. If in one possible world, someone makes certain speech sounds, and if in another possible world, this same person makes the very same sounds, we are not forced to conclude that the very same open-sentence token was produced. If, for some reason, the sounds produced in the second possible world have a different meaning or refer to different things, we can stipulate that a different open-sentence token was produced. The idea is to stipulate that if S and T have different extensions in the actual world, then they simply cannot be the same open-sentence, regardless of how much they may appear to be the same. For if the meaning of the words or the referential mechanism involved in determining the extensions of the two open-sentence tokens are so different in w_1 and w_2 as to yield different extensions with respect to the actual world, then it would be reasonable to conclude that the open-sentence tokens are different tokens. That is how we are to use the term 'open-sentence token' in this model.

The above considerations suggest a somewhat different way of avoiding the sort of problems being discussed. One can make use of a feature of counterpart theory—a theory of the semantics of modality introduced by David Lewis in (Counterpart). The basic idea of counterpart theory is to bypass problems about transworld identity, by taking each possible thing to be tied to a world, so that nothing exists in more than one world (Counterpart, p. 126). Semantically speaking, this comes down to the requirement that if world $\mathbf{w}_1$ is different from $\mathbf{w}_2$, then $U(\mathbf{w}_1)$ and $U(\mathbf{w}_2)$ are disjoint. Thus, there will be no problem, in this kind of semantical system, of trying to make sense of an object in one world being identical to an object in another world. But although nothing can exist in more than one world, a thing in one world can have *counterparts* in other worlds. The view is that if, say, Steven Sondheim has a counterpart X in some other possible world $\mathbf{w}$, then X will resemble Steven Sondheim more than anything else in $\mathbf{w}$. By making use of the counterpart relation in place of transworld identity, counterpart-theorists are able to construct a promising analysis of modal notions.

There is no need, for my purposes, however, to go into the details of counterpart theory.[2] The point is: we can stipulate that the open-sentences talked about in **Ct** are to be "world-bound individuals", as are the things in counterpart-theoretic semantics. Since no open-sentence will appear in more than one world, we cannot have the troublesome sort of situation considered earlier, in which an open-sentence in one world satisfies X, while that same open-sentence in another world does not satisfy X. Similarly, we cannot have the situation described above in which an open-sentence in $\mathbf{w}_1$ is satisfied by some object a, while this same open-sentence appearing in another world $\mathbf{w}_2$ is not satisfied by a.

Will this requirement that open-sentences be world-bound necessitate the introduction of a counterpart relation for **Ct**? Not at all. Counterparts were introduced in order to have valuations of sentences at different possible worlds without transworld identity. But since the predicates of **Ct** are all rigid, there is no need to evaluate the truth-values of sentences of **Ct** at different possible worlds.

I turn now to the axioms of **Ct**.

THE AXIOM OF EXTENSIONALITY

What this axiom says can be seen to be rather trivial when one has in mind the intended interpretation of the I-predicates. Thus, suppose F and G are different open-sentence tokens of level 1 and suppose that F is extensionally identical to G, i.e. suppose that every object that

---

[2] For details, see D. Lewis (Counterpart), A. Hazen (C Semantics), and G. Forbes (Modality), ch. 3, sects. 4 and 5.

satisfies F satisfies G, and conversely. Then, according to this axiom, F is extensionally identical to G, i.e. F and G are in the I relation. So the axiom only pins down how we wish the I-predicate to be interpreted. I give below the axiom for the cases in which n is greater than 0. The level-0 case differs only in the kind of quantifier to be used in the statement of the axiom.

If α and β are variables of level $n + 1$, γ is a variable of level n, then

$$(A\alpha)(A\beta)(A\gamma)((S^n\gamma\alpha \longleftrightarrow S^n\gamma\beta) \to I^{n+1}\alpha\beta)$$

is an axiom of extensionality.

IDENTITY AXIOMS

These axioms allow us to use the I-predicate as we would the identity predicate in set theory. This may seem paradoxical since extensional identity cannot in general be expected to behave in the way strict identity does. So a justification is needed for these axioms. This will be given below after the axioms are given.

1. If α is a variable of level 0, then

$$(\alpha)I\alpha\alpha$$

is an identity axiom.

2. If α is a variable of level n ($n > 0$), then

$$(A\alpha)I^n\alpha\alpha$$

is an identity axiom.

3. If α and β are variables of level n and if φ and ψ are formulas in which neither α nor β occur bound, then if ψ is exactly like φ except that α and β have been exchanged at one or more places, any universal closure of

$$I^n\alpha\beta \to (\varphi \longleftrightarrow \psi)$$

is an identity axiom.

Turning now to the justification of these axioms, it is clear that, given the intended interpretation, cases (1) and (2) must be true. But (3) seems to pose a serious problem. Let us consider a simple case. Suppose that F and G are open-sentences of the same level and that they are extensionally identical. Suppose further that F satisfies some open-sentence **H** of the next level. Then according to (3) above, *G* must also satisfy **H**. But why should that be?

Let us say that a level *n* + 1 open-sentence **H** is *extensional over* the level-*n* open-sentences if, and only if, for all open-sentences *F* and *G* of level *n*, if *F* is extensionally identical to *G* and **H***F*, then **H***G*. Now we can proceed to prune the totality of open-sentences we consider in the theory, as we did in the previous chapter. Thus, begin at level 2. Every open-sentence of level 2 is either extensional over the level-1 open-sentences or it is not. Let us consider in this theory only those that are extensional over the level-1 open-sentences. We can make a similar restriction at level 3: we only allow into the range of the variables those open-sentences of level 3 that are extensional over the open-sentences that are in the range of the level-2 variables. Similarly for levels 4, 5, 6, and so on. In other words, we restrict the open-sentences to be discussed in the theory to those that are extensional over their respective argument ranges. Thus, the constructibility quantifier '(**C***x*)' is to be understood as saying 'It is possible to construct a well-defined level-*n* open sentence *x* that is extensional over the well-defined level (*n* − 1) open-sentences that are extensional over the well-defined level (*n* − 2) open-sentences that are . . . such that . . .'.

The above passage needs clarification evidently, since the following kind of objection has been made to me several times.[3]

> The above passage makes it seem as if the different levels could be pruned successively: as if the open-sentences to be pruned at the *n*[th] level could be simply characterized in terms of the open-sentences that had been pruned before the *n*th level. But that is not how it would work. The problem is that a level-2 open-sentence may contain a level-3 bound variable, and which level-1 open-sentence the level-2 open-sentence is true of may depend on what level-3 open-sentences the bound variable ranges over. When we first start pruning away at the level-2 open-sentences, we are still thinking of the higher-level variables as ranging over all well-defined open-sentences of the appropriate levels. On the basis of this interpretation of the higher-level variables, we determine which first-level open-sentence a given second-level open-sentence is true of, and so we determine which second-level open-sentences are extensional over the first-level open-sentences. This, in turn, will change our

---

[3] By both Jack Silver and Vann McGee.

judgements of which third-level open-sentences are extensional over the second-level open-sentences, which will force us again to revise our judgements of which first-level open-sentences satisfy which second-level open-sentences and which second-level open-sentences are extensional over the first-level open-sentences. In this way, we get hopelessly bogged down, sinking deeper the more we struggle frantically to escape.

In order to see what is wrong with this objection, it is important to distinguish clearly between the open-sentences of the language of **Ct**, that is, the language **Lt** interpreted in the way specified above, and the open-sentences talked about by this language. The process of pruning described above is the way in which restrictions are placed upon the ranges of the constructibility quantifiers of the language of **Ct**; it does not place any restrictions on the quantifiers of the open-sentences talked about by this language. To see this point more clearly, let us examine the hierarchy of possible open-sentences discussed above. We start with an unpruned hierarchy H. If the argument variable of an open-sentence F ranges over things of level n, then F is in level $(n+1)$; and if each open-sentence of level n either satisfies F or does not satisfy F, then F is well-defined over the level-n entities and goes into the next hierarchy H^{wd}. Thus, we prune H of all open-sentences that are not well-defined over their argument ranges to obtain H^{wd}. Let us now turn to the level-2 open-sentences in H^{wd}. Some of these open-sentences are extensional over the level-1 open-sentences of the hierarchy; some are not. Now in the language of **Ct**, the level-2 constructibility quantifiers are restricted to those that are extensional over the level-1 open-sentences. Similarly, the level-3 constructibility quantifiers are restricted to those open-sentences of level 3 (of H^{wd}) that are extensional over the level-2 open-sentences of the hierarchy. And so on for all levels higher. In other words, we prune away from H^{wd} all open-sentences of level 2 or higher that are not extensional over their argument ranges. This gives us the new hierarchy H^e, which is the one used in the language of **Ct**. It can be seen that the appearance of vicious-circularity arises from either the belief that the open-sentences being talked about in the language of **Ct** are only open-sentences of this very language or the belief that any restrictions I make on the ranges of the constructibility quantifiers of this language become restrictions on the constructibility quantifiers of any open-sentence that is in the range of any of these quantifiers.

But what are the consequences of restricting, in the way described above, the constructibility quantifiers of the language of **Ct**? Might we not thereby eliminate too many possible open-sentences from consideration by the theory? One way of responding to such a question is to note that the restriction to open-sentences "extensional over their argument ranges" can be seen to be the pruning away of only unwanted structure, leaving intact the intended mathematical structure. But to investigate further the question of whether we have eliminated from consideration by the theory too many open-sentences, let us turn to the next axiom.

THE AXIOM OF ABSTRACTION

If α and β are variables of level n and $(n+1)$ respectively $(n > 0)$ and

$$\ldots \alpha \ldots$$

stands in place of a formula containing a free occurrence of α but not of β, then any universal closure of

$$(C\beta)\,(A\alpha)\,(S^n\alpha\beta \longleftrightarrow \ldots \alpha \ldots)$$

is an abstraction axiom. If α and β are variables of level 0 and 1 respectively, then any universal closure of

$$(C\beta)\,(\alpha)\,(S\alpha\beta \longleftrightarrow \ldots \alpha \ldots)$$

is an abstraction axiom.

To see how this axiom is to be justified, consider first the lowest-level case, in which S is the S-predicate of level 0. Let us consider the case in which the formula expressing the condition contains no free occurrences of any variable other than α. In that case, the axiom asserts that it is possible to construct a well-defined first-level open-sentence that is satisfied by just those objects that satisfy the condition in question. But clearly it is possible to construct such an open-sentence since it is possible to construct the formula expressing the condition, i.e. the formula expressing the condition is itself an open-sentence of the required sort.

Consider next the case in which the formula expressing the condition contains free occurrences of just one variable, δ, other than

α and that this variable is of level 0. The question is: no matter what value δ may take, will it be possible to construct an open-sentence of the appropriate sort that is satisfied by just those objects that satisfy the resulting condition? Suppose that δ takes as value the object k. Then it is reasonable to maintain that there is some possible world in which the language of this theory is extended to include a name of the object k. Then the formula expressing the condition in question can be converted into the required open-sentence by replacing all free occurrences of δ in this formula by the name of k, and surely it is possible to do this.

Finally, consider the case in which only one higher-level variable δ occurs free in the formula expressing the condition. Then when this variable takes as value some open-sentence token, say F, we have a condition on α. But again, it is surely possible that in some possible world the language of this theory is extended to include a name of F. Then again, the formula expressing the condition can be converted into the required open-sentence by replacing all free occurrences of δ with that name. And surely it would be possible to construct such a formula. It needs to be mentioned that in order to make the above-mentioned conversion, it is not necessary that F be regarded as an open-sentence of the language of the formula; nor is it necessary that there be in that language the means of expressing what F does. All that is required is that this language contain a name of F.

Clearly, the cases in which any finite number of variables occur free in the formula expressing the condition can be similarly shown to be true under the intended interpretation. So we can turn to the higher-level instances of the axiom. Consider

$$(\mathbf{C}y^{n+1})(x^n)(\mathbf{S}^n x^n y^{n+1} \longleftrightarrow \ldots x^n \ldots)$$

Here, the only truly new worry we have that we didn't have for the lowest-level case is that the formula expressing the condition $\ldots x^n$ $\ldots$ may not be extensional over the level-n open-sentences. For the Abstraction Axiom asserts the possibility of constructing an open-sentence token that is both well-defined *and extensional* over the level-n open-sentences; so if one is to argue in the way I did for the lowest-level case, one needs to provide grounds for maintaining that the formula expressing the condition $\ldots x^n \ldots$ is indeed extensional over the level-n open-sentences. To show this, I first define the *index of a formula* to be the number of occurrences in the formula of connectives and/or quantifiers. I then proceed by induction on the index of formulas to show the following:

> If φ is an open-sentence token of a formula (of **Lt**) in which only α, a variable of level n, occurs free, and if **Lt** is given the intended interpretation, then, no matter what open-sentence tokens are assigned to the individual constants of **Lt** so long as the tokens are all from the appropriate level argument ranges, φ is extensional over the level-n open-sentences.

1. Suppose φ is a formula of index 0. Then φ is atomic. I shall consider only the two cases involving the appropriate S-predicate since the ones involving the I-predicate are so trivial. Thus, consider first:

[1] $\varphi = S^n \alpha \beta$.

Then β denotes some level $(n+1)$ open-sentence token which is extensional over the level-n open-sentences. Suppose that F and G are open-sentence tokens in the range of α, that F satisfies φ, and that G is extensionally identical to F. Since F satisfies φ, according to the intended interpretation F must satisfy the token open-sentence denoted by β. But that token, i.e. the one denoted by β, must be extensional over the level-n open-sentences; so G must satisfy it. It follows that G must satisfy φ.

[2] $\varphi = S^n \beta \alpha$.

Then β denotes some level-n open-sentence token. Suppose that F and G are open-sentence tokens in the range of α, that F satisfies φ, and that G is extensionally identical to F. Then, given the intended interpretation of **Lt**, the token denoted by β must satisfy F. And since G is extensionally identical to F, that token denoted by β must satisfy G. Hence, G satisfies φ.

2. I now adopt the inductive hypothesis that what we wish to show above for all formulas holds for all formulas of index less than k. Suppose that φ is a formula of index k. I wish to show that the above holds for φ. To do this, I note that φ must be of one of the following nine forms:

[1] $\varphi = -\theta$.
[2] $\varphi = (\theta \vee \psi)$.
[3] $\varphi = (\theta \& \psi)$.
[4] $\varphi = (\theta \rightarrow \psi)$.
[5] $\varphi = (\theta \longleftrightarrow \psi)$.

[6] $\varphi = (\mathbf{C}\alpha)\theta$.

[7] $\varphi = (\mathbf{A}\alpha)\theta$.

[8] $\varphi = (\exists\alpha)\theta$.

[9] $\varphi = (\alpha)\theta$.

Consider case [1]. Notice that θ must have index less than k. Hence it must be extensional over the level-k open-sentences. Hence φ must be also. Since all the other connective cases [2] through [5] are equally simple, I turn to the quantifier cases.

Consider [6]. Suppose that F and G are open-sentence tokens of level n, that F satisfies φ, and that G is extensionally identical to F. Since F satisfies φ, it must be possible to construct an open-sentence token H of the level of α which is such that taking β, the first individual constant of the level of α that does not occur in β, to denote H, and keeping everything else in the interpretation fixed, $\theta\alpha/\beta$ is satisfied by F. But $\theta\alpha/\beta$ is a formula of index $k - 1$. So by the inductive hypothesis, $\theta\alpha/\beta$ is extensional over the level-n open-sentences, even when β denotes H. Thus, $\theta\alpha/\beta$ is satisfied by G. Thus, G satisfies φ.

The other quantifier cases [7] through [9] can be shown in an analogous fashion.

THE HYPOTHESIS OF INFINITY

In order to develop the arithmetic of natural numbers and the basic framework for the cardinality theory of Simple Type Theory, I need to appeal to a sort of hypothesis of the system. Generally, what I have in mind is called an 'Axiom of Infinity', but for reasons I shall give shortly, I prefer to call it the 'Hypothesis of Infinity'. Basically, this hypothesis says that there are infinitely many "objects" or level-0 entities. In abbreviated form, it can be expressed by the sentence

$$- (\mathbf{C}y^2)(\mathrm{Nn}(y^2) \ \& \ (\mathbf{C}x^1)(S^1x^1y^2 \ \& \ (z)Szx^1))$$

where 'Nn ( )' is to be taken as an abbreviation of a formula of the system which defines the level-3 "property" of being a level-2 natural number (or finite cardinality) "property". Thus, a formal statement of this hypothesis requires the development of a sufficient amount of finite cardinality theory to allow the expression of the above-mentioned level-3 "property". Intuitively speaking, a level-2 open-sentence F has the "property" expressed by Nn if, and only if, F is extensionally identical to either an open-sentence expressing the "property" of having no objects falling under it or an open-sentence expressing the "property" of having one object falling under it, or an open-sentence expressing the "property" of having two objects falling under it, or . . . (How Nn is to be defined formally in the present system will be discussed in the following chapter.) It can be seen that the Hypothesis of Infinity asserts that no level-1 "property" possessed by all the objects in the domain has finite cardinality.

I have been speaking up to now of how we are to interpret the language Lt and the intended interpretation of the language. But in a way, speaking of "the intended interpretation" is misleading, for I have not specified how we are to understand the level-0 individual symbols. What are the "objects" that I have been talking about? The classical Logicists, Frege and Russell, thought that there was some ontologically (or logically) basic totality—"objects" for Frege and "individuals" for Russell—that the lowest-level variables were supposed to range over. I believe there is general (or, at least, widespread) agreement among Anglo-American philosophers today that the metaphysical views of the above classical Logicians on this issue are questionable. But there is less agreement when Quine's ideas on this question are considered. I have in mind, in particular, Quine's tendency to presuppose some such basic totality, as when he interprets the logical sentence '$(\exists x)Fx$' to mean 'there is an entity x that is F'. (For example, in (Reification), p. 102, Quine writes: "The quantifiers '$(\exists x)$' and '(x)' mean 'There is some entity x such that' and 'each entity x is such that'", as if that tells us what the domain of the interpretation is.) When the existential quantifier is so understood, the bound variable can be regarded as ranging over the totality of *entities*. But what are "entities"? Unfortunately, Quine never bothered to make clear, to my mind at least, just what these "entities" are supposed to be. Consider the sentence

$(\exists x)(x$ is a message coming over the telegraph wire now$)$

and imagine that we "see" a message coming over the telegraph. Is the above sentence true when the existential quantifier is understood in the way Quine suggests? I, for one, cannot tell, since I don't know how to determine whether to classify the sequence of clicks being produced by the movement of the telegraph key as an "entity". Is the state of California an entity? Are events? Are rainbows? How do we tell?

Might not Quine respond to the above by defining the term

'entities' as "whatever satisfies the formula '$x = x$'" or as "just those items that are self-identical"?[4] But does this "definition" really help? After all, Frege could have given essentially the same definition of 'object'; for in Frege's system, objects are the only things that satisfy '$x = x$'. For Frege, strict identity is a relation that applies only to objects. Should we conclude then that Quine's entities are the same as Fregean objects? I doubt that Quine would want us to draw such a conclusion. After all, Frege's system of logic makes assertions that begin with existential quantification over concepts—things that cannot be identified with his objects.

Consider the question 'Are rainbows entities?' The above "definition" tells us that rainbows are entities if, and only if, they are self-identical. But how do we tell if rainbows are self-identical? Or consider the case of nations. Do nations satisfy the condition '$x = x$'? In so far as I have any idea of what Quine was thinking of when he spoke of entities, I would suppose that nations are not entities. But the above "definition" in terms of '$x = x$' doesn't help me at all in determining this.

Another possibility that has been suggested to me is to define 'entity' as follows: entities are simply what there are. In other words, x is an entity if, and only if, x exists. Does this help? Consider the message example again. Is the sequence of clicks being produced by the movement of the telegraph key an entity? The above definition tells us that it is if, and only if, it exists. Well, does it exist?

It might be said that I am only pointing to a difficulty in *deciding* what things truly exist. Such difficulties in no way show that the question of whether or not the sequence of clicks is an entity is not well-defined or is not a factual question. I grant that. I do not know if that question is difficult to answer because it is difficult to determine if the sequence of clicks truly exists or because it is not clear what it even means to say that the sequence of clicks exists (or is an entity). Suffice it to say that it is by no means obvious to me that the question we are considering is a coherent, but difficult to answer, question of fact.

---

[4] This was suggested to me by a referee for *Journal of Philosophical Logic*. On the other hand, I detect a kindred soul on this issue in the writings of Robert Stalnaker. Cf. "The concept [of possible world] is a formal or functional notion, like the notion of an *individual* presupposed by the semantics for extensional quantification theory. An individual is not a particular kind of thing; it is a particular role that things of any kind may occupy: the role of subject of predication. To accept the semantics for quantification theory is not to accept any particular metaphysics of individuals" (Inquiry, p. 57).

I wish to emphasize that I have not been trying to show that the sort of basic domain of quantification the Classical Logicists appealed to makes no sense or that the term 'entity', as Quine used it, is meaningless or defective. Far from it. I only wish to indicate some reasons for *my reluctance to assume such domains in my theorizing* here and for restricting quantification in **Ct** to more manageable totalities. My own view is that the notion of *thing* or *entity* is context-dependent: something to be clarified or specified, depending on what is to be discussed, analysed, or thought about. To avoid problems having to do with what entities, objects, or individuals are, I prefer to work with a system which permits a variety of different choices of domains of objects, depending on the purposes for which the system is to be used. For the analyses of many everyday uses of cardinality theory, the domain of objects can be restricted to relatively small macroscopic objects that can be easily lifted by a single adult. (In analysing the statement 'There are five apples on the table', it is not necessary to concern ourselves with questions about the nature of Quinian entities.) Thus, when I speak in this work of the 'intended interpretation' of **Lt**, it should be understood that I have in mind only a partial interpretation of the language, and that the choice of the domain of level-0 objects has not been made.

It should now be clear why I do not take what I have called the 'Hypothesis of Infinity' to be an axiom—something that is true under the intended interpretation. I cannot suppose, as did Russell, that the question of whether the domain of objects is finite or not is simply a matter of fact, to be settled, if at all, by the appropriate scientific investigation. Nor is there any reason to maintain that the hypothesis will hold no matter what domain of objects we may select, since I want to allow interpretations in which the domain of objects is finite. Thus, when finite cardinality theory is viewed from the perspective of the present system, it will appear to be concerned with possible situations in which the following holds: for every finite cardinal "property" (that of *having cardinality n*), there corresponds an extensionally different succeeding finite cardinality "property" (that of *having cardinality $(n+1)$*). In other words, finite cardinality theory will be concerned with situations in which the Hypothesis of Infinity holds. Thus, we can regard the number theorist as implicitly adopting the Hypothesis of Infinity as an axiom. But in the present system, the hypothesis will function as merely a hypothesis: certain theorems will presuppose the hypothesis and others will not. The ones that do can be converted into a theorem of the system by conditionalizing on the hypothesis (using rule (C)).

But can we find an interpretation of the system that would clearly satisfy the Hypothesis of Infinity? After all, we need to be assured that the Simple Type theoretical version of Cantorean set theory will come out true under some interpretation of the system: we shall want to have at least one way of interpreting **Ct** which "preserves" the basic structure of classical mathematics. One way of obtaining such an interpretation is to make use of some effective rule for generating an infinity of strings of symbols from some finite vocabulary, as for example, the simple rule we have all learned as children for producing the counting numerals in the Arabic notation. Call that rule '*R*'. We then shall treat the level-zero quantifiers as constructibility quantifiers. In particular, the level-zero existential quantifier '$(\exists x)$' will be construed as saying:

> It is possible to construct a counting numeral x in accordance with R such that . . .

We shall then need to find an appropriate way of interpreting the level-zero I-predicate. Strict identity would seem to be the natural choice, but such a choice could lead to philosophical problems analogous to those having to do with transworld identity, i.e. strict identity between things in different possible worlds. To avoid any such problems, let us consider using an equivalence relation that is "weaker" than strict identity. Now any counting numeral constructed in accordance with R will be a string of Arabic numerals. So let us use the idea of one counting numeral being a string of the same type as another counting numeral. We can stipulate that a counting numeral N constructed in accordance with R is *the same type of numerical string as* counting numeral M constructed in accordance with R if, and only if,

> the leftmost symbol of N is the same (type of) Arabic numeral as the leftmost symbol of M; and

> the next symbol of N is the same (type of) Arabic numeral as the next symbol of M;

> etc.

We can then interpret the zero-level I-predicate to be given by:

is the same type of numerical string as

It can then be seen that the Hypothesis of Infinity will be true under the interpretation being sketched here.

It can also be mentioned that in a later chapter (Chapter 6), I shall describe an idealized state of affairs in which the objects described satisfy the Hypothesis of Infinity.

I shall conclude this section with a few remarks comparing my version of type theory with Russell's. Both systems are, in some sense, "no-class" theories: in both systems, there is no appeal to classes or sets. However, in Russell's system, set theory is developed by means of translation rules, which allow one to construct sentences that are apparently about sets but that are, in fact, merely abbreviated versions of sentences that refer only to individuals and propositional functions. (See my (V-C), ch. 1, sect. 4 for details.) Russell's translation rules are cumbersome, and even the simplest of set-theoretical sentences, according to this analysis, can be riddled with ambiguities. My own approach is to forgo even apparent reference to sets. Instead, one talks about the constructibility of open-sentences that function like sets. The main feature of sets that needs to be captured is extensionality; and this is accomplished by restricting talk of open-sentences of level $(n + 1)$ to only those open-sentences of level $(n + 1)$ that are extensional over the level-n open-sentences.

3. Objections to Type Theory

The constructibility theory I have presented above can be characterized as a version of Simple Type Theory. In this section, I would like to consider some objections that have been raised to Simple Type Theory. It should be noted that critics of Simple Type Theory generally have in mind set-theoretical versions of Type Theory, that is, versions in which the variables are interpreted as ranging over sets or classes. Now one striking feature of such versions of Type Theory is existence of infinitely many different null classes, one for each type. It is in response to this feature of Type Theory that Quine once wrote:

One especially unnatural and awkward effect of the type theory is the infinite reduplication of each logically definable class. . . . This reduplication is particularly strange in the case of the null class. One feels that classes should differ only with respect to their members, and this is obviously not true of the various null classes (Types, p. 131).

I agree that there is something absurd about the notion of different null classes. A class X is supposed to be identical to class Y if, and only if, X and Y have the same members. But what distinguishes the null class of type 0 from the null class of type 1? It would seem that what distinguishes these two classes is not what members they have, but rather what members they do not have: the former class has no member of type 0, whereas the latter has no member of type 1. This reminds me of the story of the man in a café who asks for coffee without cream and is told, "Sorry, we don't have any cream—would coffee without milk do?"

But this objection does not apply to the version of Type Theory developed in this chapter. For it is not asserted in this theory that there are different null sets: what corresponds, in this theory, to the assertion of the existence of null classes of different types is the assertion of the possibility of constructing open-sentences of different types that are not true of anything. And there is nothing absurd about being able to construct such null open-sentences—such open-sentences would differ in kind from each other; yet they would each be true of nothing.

Another objection to Simple Type Theory has been raised by J. Moss, who complains that "it is difficult to believe in the infinite reduplication of the natural numbers, and everything constructed from them, to which the theory is committed" (Sprouts, p. 233). Again, this objection does not apply to the version of Type Theory being discussed here; for in this system there is no infinite reduplication of the natural numbers. What one has, in effect, is a sort of reduplication of the structure of the natural numbers. And that is not at all absurd!

4. Objections to the Constructibility Theory

Before turning to the development of mathematical theory within this system, I would like to consider an objection to the interpretation I have constructed here. This objection is that the constructibility interpretation is not essentially different from the model-theoretic interpretations of Simple Type Theory based on the Löwenheim–

Skolem Theorem, and that since it is generally agreed that these so-called "Skolem reductions" are unsatisfactory, it should also be agreed that the constructibility interpretation I give here is also unsatisfactory.

Consider again Euclid's constructibility theory. Now we can distinguish two sorts of interpretations of formal systems of geometry: (1) the Platonic kind according to which the axioms postulate the existence of certain abstract entities (points, lines, planes, and the like); and (2) Euclid's according to which the axioms make assertions about *what it is possible to construct*. These kinds of interpretations are *independent* in the sense that each can be comprehended without appeal to the other. Thus, Euclid's system was not specified in terms of the Platonic. Historically, Euclid's system came first. Now I would claim that the constructibility interpretation I give here is independent (in this sense) of the various Platonic interpretations of the formal system that one could construct. It is perfectly conceivable that someone could have produced the formal system described in this chapter, and given it the constructibility interpretation, before set-theoretical versions of Simple Type Theory were developed.

What about "Skolem reductions"? These are supposed to be reductions that make use of the theorem that says, roughly, if a formal theory has a model, it has a denumerable model. Given some Platonic theory, then, we can interpret it as being about the denumerable model proved to exist by the Löwenheim–Skolem theorem. But these "Skolem reductions" of Platonic theories are generally parasitic in some way or other on the Platonic model being reduced. Thus, if the denumerable model is obtained by operating on the Platonic model itself, we get an unacceptable situation in which we presuppose the very thing we want to reject. So let us consider the Henkin proof of the Löwenheim–Skolem theorem, in which the denumerable interpretation is specified not in terms of the model, but rather in terms of the *sentences* of the theory that are asserted. If we give such a Skolem interpretation to, say Zermelo–Fraenkel set theory, then the ϵ-relation would be specified, in effect, in terms of the assertions of the theory. For definiteness, let us take the domain of the Skolem interpretation to be a denumerable set of tokens. Then 'ϵ' is interpreted to be a complicated relation among the totality of these tokens. But to apply this theory, to the world, we need more than just the set-theoretical axioms. In addition to the mathematical truths, we are going to need empirical ones, as well as truths relating empirical objects with mathematical ones. But by the Henkin method, how 'ϵ' gets interpreted is dependent on the set of assertions of the theory being interpreted. Since we are continually changing our set of assertions (we are continually discovering new things, revising our theories, redistributing truth-values among the sentences of the theory), this would require a continual revision of our interpretation of 'ϵ'—which certainly points to the sham quality of this reduction. It is clear that the Skolem reducer does not really take seriously the interpretation given. She does not *interpret* the theory in any ordinary sense of that term. For the Skolem interpretation does not assign any real *meaning* to the primitive terms: one cannot go on to discover new truths about the entities in the domain by making observations and expressing what one learns about these entities in terms of the relations specified by the Skolem interpretation. On the contrary, the sentences to be rejected or accepted are determined without attention to the tokens supposedly being talked about, and then the Skolem reducer keeps adjusting the Skolem interpretation so that one still has a model of the revised theory.

Is this the way one must proceed with the constructibility interpretation being presented here? Not at all. As new empirical data are gathered, the set of truths of a scientific theory utilizing the constructibility system would change. But the meaning given to the S-predicates and the I-predicates of **Ct** will not change as a result: the S-predicates will continue to mean satisfaction and the I-predicates will continue to mean identity or extensional identity (depending on the level), regardless of the addition of new sentences to the body of accepted ones. Nor is there any need to see what truths would be accepted according to some Platonic interpretation of the system to see what changes should be made in the constructibility interpretation, for the constructibility interpretation is not parasitic on such Platonic interpretations.

The above Skolem interpretation provides us with a rather mysterious relation among tokens of symbols. Why a theory about this relation should prove to be so useful in the empirical sciences and in practical matters would be a puzzle for the Skolem reducer. The question would arise: What justifies us in applying the mathematical theory in the empirical sciences to draw inferences about genes, chemical processes, black body radiation, and the like? Well, is there not a similar puzzle about a theory that seems to give us only information about the constructibility of open-sentences? How, it might be asked, can such a theory be used to yield any useful information about the physical world?

Consider the following example. Suppose that certain sentences of a first-order language $A_1, A_2, \ldots, A_n$ are established to be true by the empirical scientist. Then, the information that it is possible to construct a list of sentences $B_1, B_2, \ldots, B_m$ which satisfy the syntactical rules of being a derivation of B_m from $\{A_1, \ldots, A_n\}$ would enable us to infer the truth of B_m. It is not at all hard to see how this information about the constructibility of sentence tokens could prove to be enormously useful.

So now, take the case of cardinality. How can a theory of the constructibility of open-sentences enable us to infer how many sheep someone owns, given information about how many sheep the person owns in various locations? How can such a theory aid us in drawing inferences about the lengths, volumes, and masses of physical objects? The following chapters will provide some detailed and specific information relevant to these questions. However, even before any specific analysis of cardinality is given from the perspective of the constructibility theory, one can see, in a general way, how information about the constructibility of open-sentences can be of central importance in dealing with cardinality. For it can be seen that the logical work done by the existence of a set which is a one-one relation can be done by the constructibility of an open-sentence that correlates one-one.

The above considerations are relevant to an objection to "classical nominalism" raised by Philip Kitcher, who claims that the classical nominalist "faces the puzzle of why studying the properties of physical inscriptions should be of interest and of service to us in coping with nature" (Nature, p. 115). He then goes on to say that he rejects nominalism because "we do not receive an adequate account of the usefulness of arithmetic". It may be thought that this objection can be carried over to the constructibility theory being presented here. After all, is not the constructibility theory a theory about open-sentence tokens? And are not open-sentence tokens things that can be produced by constructing "physical inscriptions"? It needs to be emphasized that the constructibility theory is not concerned with *the physical properties of physical inscriptions*. Nor is the constructibility theory concerned with actually existing physical inscriptions. How a theory about what open-sentences are constructible can be useful in coping with nature will be discussed in detail in the following chapters.

I emphasize the above ontological point because my position tends to be misunderstood or misdescribed. For example, Field has objected to my earlier predicative type-theoretical approach by attributing to me an appeal to "sentence types no token of which has ever been uttered" (Field (Science), p. 45), suggesting that my theory implicitly refers to, and thus assumes the existence of, abstract entities. The sentences of my theory do assert the possibility of constructing open-sentences of certain sorts. But this is not to *refer* to any open-sentences, actual or possible. If I say that it is possible to construct a playground next to my house, I am not referring to, or presupposing the existence of, a possible playground. In this sense, I am not "appealing" to a playground that has never been constructed.

Field has also written, ". . . physics, as I see it, has as its only concern to describe the actual features of the actual world and I don't think it appropriate to invoke facts about *what is possible* in doing this" (Field (Revision), p. 282, italics mine). I ask the reader to keep an open mind about the potential utility and reasonableness of theorizing in terms of what it is possible to construct—whether in physics or in everyday life. It seems to me that these are not matters to be resolved by intuitions or hunches about what is appropriate in physics: they are matters for serious investigation and study.

Let us consider one last objection which was suggested to me by Hans Sluga. Recall that the semantics of the language **Lt** was given in terms of sets. But how can I accept such a semantical analysis, given that I am sceptical about there being any such things as sets?

I have already indicated (in Chapter 2) one sort of reply to this objection. One can regard the constructibility quantifier as a primitive of the system and view the possible-worlds semantical models as mere didactic devices, useful in communicating my theory to those who are familiar with traditional semantical theories. Aleksandar Ignjatović has suggested to me another way of understanding what I have done. The model theory I use in this chapter can be easily expressed in Simple Type Theory; so this model theory can be developed within the constructibility theory itself. Thus, I do not have to presuppose the existence of sets to accept the semantical account presented earlier.

Some may wonder about the legitimacy of proceeding in this way. It may seem strange to use a model theory to explain the logical significance of various elements of the constructibility theory, while at the same time developing this model theory within the constructibility theory itself. But such a procedure is really not so unusual. A set theorist, for example, may use a model theory to explain the logical significance of various elements of her set-theoretical language, while at the same time developing this model theory within the set theory itself.

5

Cardinality and Number Theory

I NOW turn to an analysis of the mathematical reasoning that can be developed within the logical framework we have been considering. Let us begin with the fundamentals of cardinality theory. Although all the basic ideas to be presented here can be traced back to Frege's (Foundations), the theory of cardinality I shall sketch will of necessity differ from Frege's because of differences in the respective logical systems used. I shall be concerned in this chapter with developing the fundamentals of the theory of finite cardinality of open-sentence tokens of level 1, even though the theory easily generalizes to the higher levels. Thus, it will be these level-1 open-sentence tokens that will be said to have cardinality.

1. Formal developments

Some Notational Conventions

In this chapter, I shall use the term 'property' to refer to level-one monadic open-sentence tokens, and 'attribute' to refer to level-two monadic open-sentence tokens. 'Object' in what follows will always refer to an element of the actual world. Now if O is an object and F a property, O can be said to "have the property F" just in case O satisfies F. Similarly, property F will be said to have the attribute H if, and only if, F satisfies H.

I shall now suggest a method of obtaining open-sentence tokens that function as ordered pairs. Let us begin at level 3. In the following, the predicate variables (upper-case letters) can all be thought of as ranging over monadic open-sentence tokens.

> *Couples* (or unordered pairs). For all objects x and y, and any property F, F is a couple $\{x, y\}$ if, and only if, for every object z, z satisfies F just in case z is identical to x or to y.

> *Ordered Pair.* For all objects x and y, and any attribute H, H is an ordered pair $<x, y>$ if, and only if, for any property G, G satisfies H just in case G is extensionally identical to a couple $\{x, x\}$ or G is extensionally identical to a couple $\{x, y\}$.

Relations. If R is an open-sentence of level 3, then R is a relation if, and only if, all the attributes **H** that satisfy R are ordered pairs.

Comment. In effect, I have defined: couples of objects, ordered pairs of objects, and relations among objects. By an obvious extrapolation from the above, one can define: couples of properties, ordered pairs of properties, and relations among properties; couples of attributes, ordered pairs of attributes, and relations among attributes; and so on. Having defined 'ordered pair' and 'relation', it is a relatively simple task to reconstruct the Fregean theory of cardinality in this new setting.

Notation. Suppose x and y are objects and R is a relation. I use 'xRy' as an abbreviation for 'an ordered pair $<x, y>$ satisfies R'.

I now give a series of definitions. These will be given informally to facilitate grasping the main ideas behind the formal developments I have in mind.

1. *Correlation*

    A relation R *correlates* a property F with a property G if, and only if, for every object x, if Fx, then there is an object y such that Gy and xRy; and if Gx, then there is an object y such that Fy and yRx.

2. *One-one relation*

    A relation R is *one-one* if, and only if, for all objects x, y, z, if xRy and xRz, then $y = z$; and
    if xRy and zRy, then $x = z$.

3. *Equinumerosity*

    A property F is *equinumerous* with property G if, and only if, there is a relation that is one-one and that correlates F with G.

Comments. It is an elementary matter to prove in this system that equinumerosity is an equivalence relation among the totality of properties, i.e. is reflexive, symmetric, and transitive. In realistic terms, equinumerosity partitions the totality of properties into mutually exclusive and exhaustive equivalence classes.

4. *Cardinality attributes*

    A second-level monadic open-sentence **H** is a *cardinality attribute of F* so long as, for all G, **H**G if, and only if, G is equinumerous with F.

    H is a *cardinality attribute* if, and only if, there is some F such that **H** is a cardinality attribute of F.

(*Note*. It follows, again using realistic terminology, that a cardinality attribute is an attribute possessed by all and only the properties in one of the equivalence classes described above.)

Discussion. It may be helpful at this point to examine some concrete examples of the items defined above. Imagine the following: There is an apple, which I shall call 'Jack', on my office desk and that there is another apple, which I shall call 'Tom', in a drawer of this desk. No other apple is to be found in my office. On my desk is a pen, which I call 'Jill', and only one other pen is in my office. Call this other pen 'Sue'. I construct on my blackboard the level-1 open-sentences:

$$x^0 = \text{Jack}$$
$$x^0 = \text{Jack or } x^0 = \text{Jill}$$
$$x^0 = \text{Tom}$$
$$x^0 = \text{Tom or } x^0 = \text{Sue}$$

specifying that the variable 'x^0' is to range over objects, i.e. the things in level 0. Let us name these open-sentences '[1]', '[2]', '[3]', and '[4]', respectively. I then write

x^1 is extensionally identical to either [1] or [2]

specifying that the variable 'x^1' is to range over the level-1 open-sentences. Call this level-2 open-sentence 'Fido'. Fido is clearly well-defined and extensional over the level-1 open-sentences. I also construct

x^1 is extensionally identical to either [3] or [4]

which is named 'Fido-fido'. Fido-fido is also obviously well-defined and extensional over the level-1 open-sentences. Finally, I write on the blackboard

x^2 is extensionally identical to either Fido or Fido-fido

specifying that 'x^2' is to range over the level-2 open-sentences. Call this level-3 open-sentence 'Arf'. Arf is well-defined and extensional over the level-2 open-sentences; and clearly, all the level-2 open-sentences that satisfy Arf are ordered pairs—indeed, they must all be ordered pairs < Jack, Jill > and < Tom, Sue >. Hence Arf is a relation. Furthermore, it is a simple matter to verify that Arf is a one-one relation. Thus, writing

x^0 is an apple in my office

and

x^0 is a pen in my office

I construct level-1 open-sentences which I call 'A' and 'B' respectively.

Now Arf correlates A with B. It follows that A is equinumerous with B; so the cardinality of A = the cardinality of B. Intuitively speaking, it is obvious that the cardinality of A = the cardinality of B. But it needs to be shown that what expresses this intuitively obvious fact, in my system, can be established.

The above example, however, is somewhat misleading. For it may suggest to some readers that an assertion, in my system, of the possibility of constructing an open-sentence of such-and-such sort implies that it is possible for *the asserter* to construct an open-sentence of the kind in question. It needs to be emphasized that nothing of the sort is intended. Nor do I wish to suggest that the asserter (or group of asserters) can provide an effective rule for generating the open-sentences. In other words, the above example should not be taken to be typical. In the above, I was able to construct all the relevant open-sentence tokens that were being asserted to be constructible. But in many cases, I will not be able to do the constructions.

To reinforce this point, let us consider an example involving uncountable cardinality attributes. Among the theorems of **Ct** there is one that corresponds to the classical theorem "The power set of the natural numbers is equinumerous with the set of real numbers." Now this theorem implies that it is possible to construct uncountably many open-sentence tokens corresponding to the traditional ordered pairs in which the first member is a set of natural numbers and the second is a real number. But who could do that? Clearly, no one. But the theorem does not imply that it is possible for any single person to do any of these things. (Recall Chapter 3, Section 4.) Let us again make use of the Kripkean possible worlds model to clarify the situation. To say that it is possible to construct the open-sentence tokens in question is to say that, for each of the "possible open-sentence tokens", there is a possible world in which someone constructs the token. But there does not have to be any single possible world in which all these tokens are constructed. Indeed, there's no reason why there couldn't be uncountably many different worlds in which these tokens are to be found, no one world containing more than finitely many.

I now define a particular kind of cardinality attribute,

5. *Zero*

   A cardinality attribute is a *zero attribute* if and only if, it is satisfied by a property that no object has.

Comment. In this system, there are no cardinal numbers—there are only cardinality attributes. So in particular, there is no such object as zero. There are only zero attributes. Of course, all second-level open-sentences that are zero attributes are extensionally identical with each other.

6. *The Predecessor Relation*

Cardinality attribute **M** *immediately precedes* cardinality attribute **N** if, and only if, there is a property *F* and an object *x* such that:

    (i) *Fx*

    (ii) *F* has cardinality attribute **N**

and

    (iii) there is some property *G* such that *G* has cardinality attribute **M** and is satisfied by all and only those objects different from *x* that satisfy *F*.

7. *Hereditary Relations*

A level-three open-sentence ψ is *P-hereditary* if, and only if, for all cardinality attributes **M** and **N**, if **M** satisfies ψ and **M** immediately precedes **N**, then **N** satisfies ψ.

8. *Ancestors*

A cardinality attribute **M** is a *P-ancestor* of cardinality attribute **N** if, and only if, every level-three open-sentence ψ is such that if

    ψ is P-hereditary

and also

    every cardinality attribute that is immediately preceded by **M** satisfies ψ

then **N** satisfies ψ.

9. *Finite Cardinality*

A cardinality attribute **N** is a *finite cardinality attribute* if, and only if, either **N** is a zero attribute or else a zero attribute is a P-ancestor of **N**.

Comment. It can be seen that as cardinality theory is developed here, there are no natural numbers, i.e. there are no objects, as there are in Frege's system, that are designated to be natural numbers. There is no need to have, in addition to the *finite cardinality attributes*, objects that "go proxy" for the attributes, for we simply theorize directly about cardinality attributes without resorting to postulating any cardinal numbers. By extrapolating from the above definitions, one can reformulate the intuitive arithmetical theorems as sentences in the formal system sketched in the previous chapter. Then, the sentences of this system that correspond to the Peano Axioms can be proven.[1] Most of these sentences can be proved without using the Hypothesis of Infinity. Thus, it is practically a direct consequence of the definition of 'finite cardinality attribute' that every zero attribute is a finite cardinality attribute (i.e. "zero is a natural number"). To prove that no zero attribute is immediately preceded by any finite cardinality attribute ("zero is not a successor"), one can simply assume that "there is" (it is possible to construct) a zero attribute that is immediately preceded by a finite cardinality attribute and then deduce a contradiction from the consequence that there is some object that has a property that satisfies the zero attribute. The axiom of mathematical induction can also be proved almost directly from the definition of 'finite cardinality attribute'. Proofs of the other axioms are somewhat more lengthy, but still they are relatively straightforward. For example, it can be proved, again without appeal to the Hypothesis of Infinity, that "the successor relation and its inverse are single valued", i.e.

> if finite cardinality attribute **M** immediately precedes both cardinality attribute **N** and cardinality attribute **R**, then **N** and **R** are finite cardinality attributes and **N** is extensionally identical to **R**.

and

> if finite cardinality attributes **M** and **N** are not extensionally identical to each other, and if **M** immediately precedes cardinality attribute **J** and **N** immediately precedes cardinality attribute **K**, then **J** is not extensionally identical to **K**.

Finally, to prove "every natural number has a successor", i.e.

> for every finite cardinality attribute **M**, there is a finite cardinality attribute **N** which is such that **M** immediately precedes **N**;

one needs to make use of the Hypothesis of Infinity.

I return to defining specific kinds of cardinality attributes. Obviously, the definitions below are not needed for the usual development of the theory of natural numbers; but they are included here for convenience.

---

[1] In her (PhD), Phyllis Rooney carries out a detailed deduction of Peano's axioms in the system of logic obtained by subtracting extensions (and hence Axiom V) from Frege's system.

10. *Specific Cardinality Attributes*

If a zero attribute immediately precedes cardinality attribute **N**, then **N** is a *one attribute*.

If a one attribute immediately precedes cardinality attribute **N**, then **N** is a *two attribute*.

.

.

.

I now define addition for cardinality attributes.

11. *Addition*

Suppose that for some properties *F* and *G*, *F* has finite cardinality attribute **M** and *G* has finite cardinality attribute **N**, and suppose that no object satisfies both *F* and *G*. Then it can be proved that there is a finite cardinality attribute **P** that is satisfied by all properties *H* that are satisfied by all and only those objects that satisfy either *F* or *G*, and that any attribute that is satisfied by all properties *H* that are satisfied by all and only those objects that satisfy either *F* or *G* must be extensionally identical to **P**. Such a cardinality attribute is to be called an **(M + N)** *attribute*.

The usual theorems concerning addition (as for example (*a*) and (*b*) below) can then be proved.

(*a*) If **N** is a finite cardinality attribute and **Z** is a zero attribute, then any **(N + Z)** attribute is extensionally identical to **N**.

Comment. It follows that if **A** is an *n* attribute, **B** is a zero attribute, and **C** is an **(A + B)** attribute, then **C** is an *n* attribute, i.e. $n + 0 = n$.

(*b*) If **N** and **M** are finite cardinality attributes and **O** is a one attribute, then any **(N + (M + O))** attribute is extensionally identical to any **((N + M) + O)** attribute.

Using the above, and following relatively standard models, it is a simple matter to prove such straightforward theorems as '*n* immediately precedes $(n + 1)$' and '$5 + 8 = 13$', i.e.

(*c*) if **N** is a finite cardinality attribute and **O** is a one attribute, then there is a finite cardinality attribute which is also an **(N + O)** attribute and which is immediately preceded by **N**

and

(*d*) if **H** is a five attribute, **G** is an eight attribute, and **J** is an **(H + G)** attribute, then **J** is a thirteen attribute.

One can also prove the standard arithmetical laws, such as the associative and commutative laws of addition. Further developments, such as specifying an appropriate definition of multiplication and proving the standard arithmetical properties of multiplication, can be carried out in pretty much the standard ways by extrapolation from the above.

2. Philosophical Amplifications

A point that needs to be emphasized is that I do not envisage using the *language of the mathematical theory being constructed* to express (empirical) scientific theories or to express one's thoughts in ordinary workaday situations. In short, I do not accept the suggestion of some logicians that true scientific reasoning ought to be carried out in the formal logical languages studied in mathematical logic. Nor do I have any sympathy with Quine's idea (described in Chapter 1, Section 2) that one ought to formulate one's scientific views in a first-order, extensional, quantificational language. My view is that scientific reasoning, as well as ordinary common-sense reasoning, for the most part can best be carried out in the usual way it in fact is, namely in natural languages. The formal language of constructibility I employ is to be used, according to my way of thinking, primarily to reason about ordinary, scientific, and mathematical reasoning. In a sense, the logical theory I construct can be regarded as a sort of tool of the metatheory, used to theorize about ordinary, scientific, and mathematical reasoning. (This idea is developed in more detail in Chapter 12.) One might say that I use my system not to reason about the world, but to reason about our reasoning about the world.[2] Furthermore, this formal system should not be taken to be the primary means by which our thoughts about our scientific reasoning are to be expressed or formulated; this will no doubt best be done in natural languages too. I see the formal system as a sort of conceptual device to aid us in our metatheorizing, in so far as it makes clear and perspicuous certain aspects of ordinary, scientific, and mathematical

---

[2] Cf. Frege's statements in (Foundations), p. 99: "The laws of number, therefore, are not really applicable to external things; they are not laws of nature. They are, however, applicable to judgements holding good of things in the external world; they are laws of the laws of nature." Notice, however, that my statements are about how I use the theory **Ct**; I am not claiming that what are generally regarded as "the laws of arithmetic" are laws of the laws of nature.

reasoning. In emphasizing this use of the constructibility theory, however, I do not wish to leave the reader with the impression that there are not other significant uses. Another use to which **Ct** can be put will be sketched in Chapter 11.

There are two quite distinct (but not necessarily incompatible) ways of viewing ordinary arithmetic: as a body of truths and as a system of useful techniques and procedures for describing and drawing conclusions about the empirical world. Frege emphasized the former in his writings, whereas Wittgenstein frequently adopted an attitude toward mathematics that brings to mind the latter.[3] There have been philosophers who have challenged the Fregean view of arithmetic, but no one, to my knowledge, has denied that arithmetic supplies us with useful techniques and procedures for drawing empirical conclusions from empirical premises. So let us adopt, for now, the latter way of regarding mathematics. Then the following questions arise: Can it be explained why these arithmetical procedures yield reasonable inferences? Can it be explained why we are justified in adopting these procedures?

I shall discuss the above questions in a more general setting in the final chapter; here, let us consider ordinary cardinality statements of the sort that Frege called "statements of number". Consider, in particular, the statement 'There are two apples on my desk.' According to Frege, this statement of number should be understood as being an assertion about a concept, namely the concept *is an apple on my desk* (Foundations, pp. 59–67). As we shall see in Chapter 10, Penelope Maddy analyses this statement as being an assertion about a set, namely the set of apples on my desk (see (Perception), p. 180). From the point of view of the constructibility theory, however, the statement can be regarded as asserting the following:

$$(AF)(x)((Fx \longleftrightarrow x \text{ is an apple on my desk}) \rightarrow F \text{ has cardinality two}).$$

This is not to suggest that the English statement means what is displayed above, but only that the constructibility statement will "do the work" of the English statement. Notice that the constructibility statement does not presuppose that there actually be anything that

[3] Cf. his comment (in (Remarks), p. 89): "I should like to say: mathematics is a motley of techniques of proof. And upon this is based its manifold applicability and its importance." See also Rush Rhees (Continuity, pp. 226–42). A number of commentators have stressed this aspect of Wittgenstein's philosophy of mathematics. See e.g. Bernays (CoRFM, pp. 93 and 97).

has cardinality two: the statement would be true even if no one has actually constructed an open-sentence token which is satisfied by just the two apples on my desk. The truth of the cardinality statement does not require that there be something (a Fregean concept or set) to which we attribute a cardinality attribute.

To make the above discussion more concrete, let us consider how the theory in question might be used to analyse a simple example of ordinary reasoning involving the use of arithmetic:

[1] There are five dimes on table A at time t.
[2] There are eight quarters on table A at time t.
[3] A coin is on table A at time t if, and only if, it is either a dime or a quarter on table A at time t.
[4] Nothing is both a dime on table A at time t and a quarter on table A at time t.

Therefore, there are thirteen coins on table A at time t.

The kind of analysis that shall be provided here should be contrasted with what has been suggested recently by Michael Detlefsen, who conceives of arithmetic as "an idealization of our experience regarding the arithmetical behavior of actual physical objects". Detlefsen asserts that:

[T]he arithmetic behaviour of physical objects is not invariant under all situations. If I take five rabbits from one pen and three from another and put them all into one pen, then, if I count the rabbits directly after putting them into the common pen, I will normally come up with eight. But if I wait six months before counting the rabbits in the combined cage, I may come up with fifteen (Hilbert's, p. 33).

Thus, according to Detlefsen, arithmetic is obtained by abstracting away those features of the physical world that make its arithmetical behaviour "unstable".

One question that should be asked, in response to Detlefsen's analysis of the rabbit example, is: What counts as unstable "arithmetical behaviour"? To put it another way, is the physical world's arithmetical behaviour unstable because the number of rabbits put into a cage does not remain constant over a long passage of time? Does arithmetic tell us that the number of rabbits should remain constant? And after we abstract away "those features of the physical world that make its arithmetical behaviour unstable", is the idealized world we obtain to be such that the number of rabbits put into a cage will always remain constant?

Detlefsen claims that "for purposes of explaining its empirical applicability, contentual arithmetic ought to be conceived as an idealization of our experience regarding the arithmetical behavior of actual physical objects" (Hilbert's, p. 33). Unfortunately, he provides his readers with no analyses of actual reasoning involving the use of arithmetic; so it is difficult to evaluate his suggestion that we can obtain plausible explanations of the use of arithmetic in everyday life and in science by conceiving of arithmetic as an idealization obtained by abstracting away features of the physical world that are arithmetically unstable.

However, Detlefsen's position is similar, in certain respects, to Philip Kitcher's view of arithmetic, as set forth in his (Nature). Kitcher presents his readers with fifteen principles, which make up the axioms of what he calls "Mill Arithmetic". (Kitcher's view of arithmetic will be discussed in detail in Chapter 11.) He then goes on to say:

Although Mill Arithmetic cannot accurately be applied to the description of the physical operation of segregation, spatial rearrangement, and so forth, that is not fatal to the applicability of Mill Arithmetic. We can conceive of the principles of Mill Arithmetic as implicit definitions of an *ideal agent.* . . . Just as we abstract from some of the accidental and complicating properties of actual gases to frame the notion of an ideal gas, so too we specify the capacities of the ideal agent by abstracting from the incidental limitations of our own collective practice (p. 117).

This is not the appropriate place for a detailed examination of these views—this will be provided in Chapter 11. Besides the obvious fact that Kitcher sees arithmetical theorems as expressing idealizations, as did Detlefsen, I note here only that, as in the case of Detlefsen, Kitcher simply does not provide his readers with any clear or detailed ideas as to how one should analyse any significant applications of mathematics or even simple everyday reasoning with cardinal numbers: neither Detlefsen nor Kitcher comes close to matching Frege's analysis of finite cardinality in this respect.

The account of finite cardinality to be provided in this chapter should also be contrasted with that of most Platonists, according to which the above arithmetical reasoning requires a belief in the existence of numbers. That such reasoning presupposes the existence of numbers has some plausibility. After all, in drawing the above conclusion, do we not make use of the theorem about $5 + 8 = 13$? And are we not using the terms '5', '8', and '13' as names of objects in the arithmetical theory of which '5 + 8 = 13' is a theorem?

To understand how this reasoning can be carried out without making reference to such mathematical objects as numbers, we need to reformulate the premises using the constructibility quantifiers. Before attempting to reformulate the premises within my system, let us first see how one would devise an appropriate interpretation of the formal language. In what follows, I shall, for simplicity, drop the type indices and make use of the informal definitions given above. It will be convenient to have in the universe of discourse, not only the objects being talked about in the reasoning, i.e. the coins on the table, but also some specific tokens of open-sentences. Thus, we can specify that '*D*' is to denote some specific token of the open-sentence '*x* is a dime on table *A* at time *t*' (it could even be the one written above). Let us use the notation

D: *x* is a dime on table *A* at time *t*

to indicate that the predicate constant '*D*' is to be interpreted as denoting the token written after the colon. Then, we can specify

Q: *x* is a quarter on table *A* at time *t*.

C: *x* is a coin on table *A* at time *t*.

Now the analysis to be given does not require that second-order predicate constants be used to name a specific open-sentence token, since they function in the analysis merely as predicates. Accordingly, for each predicate constant to be used in the analysis, I only provide an axiom that specifies the extension of the predicate. (For simplicity and perspicuity, I express the axioms in English.)

H_5 is a 5 attribute.

H_8 is an 8 attribute.

H_{5+8} is a $(H_5 + H_8)$ attribute.

H_{13} is a 13 attribute.

Then the premises of the above argument can be given in my system as:

[1*] $H_5 D$.

[2*] $H_8 Q$.

[3*] $(x)(Cx \longleftrightarrow (Dx \lor Qx))$.

[4*] $-(\exists x)(Dx \mathbin{\&} Qx)$.

It should be noted that [1*] expresses, according to Frege's analysis, what would be given by

   D has cardinality five

which would imply the *existence* of an abstract entity, a one-one relation of a certain sort. [1*], on the other hand, implies merely the possibility of constructing a one-one relating open-sentence.

From [1*], [2*], and [4*], we can infer (using the above axioms)

   [5*] $(\mathbf{A}\,Y)((x)(Yx \longleftrightarrow (Dx \vee Qx)) \rightarrow \mathbf{H}_{5+8}Y)$

so, from [3*] and [5*], we can obtain

   [6*] $\mathbf{H}_{5+8}C.$

Now we have as a theorem of the formal system something that corresponds to $5 + 8 = 13$, i.e. we have (*d*) above, so we conclude

   [7*] $\mathbf{H}_{13}C.$

Thus, we see how the formal system developed here can be used to validate in Logicist fashion aspects of everyday reasoning, such as the above, that involve finite cardinality. Furthermore, one can see by simple extrapolation from this how more complicated arithmetical theorems involving multiplication, exponentiation, factorials, and other arithmetical operations can be useful in drawing inferences about practical matters. One can also see how the system can provide essentially the Logicist analysis of the role that the relevant arithmetical theorems play in such reasoning.

Generalizing from the above analysis, one can give an account, from the point of view of the constructibility theory of cardinality sketched above, of what Frege called "statements of number", that is, simple cardinality statements of the form 'There are N *F*s'. Such a statement can be understood as saying that it is possible to construct a level-1 open-sentence that is true of all and only those things that are *F*s and this open-sentence satisfies an N attribute. In short, whereas Frege attributed cardinality attributes to level-1 concepts, in the constructibility theory, cardinality attributes are attributed to level-1 open-sentences.

By means of this theory of cardinality, questionable steps in chains of reasoning can also be isolated. As a case in point, consider an argument sometimes given to show that such simple arithmetical theorems as '$5 + 8 = 13$' are false.[4] The falsity of the theorem can be

---

[4] This argument is closely related to an example of Frege's, which was used to show that John Stuart Mill confused an application of an arithmetical proposition with the arithmetical proposition itself. See Frege (Foundations, p. 13). Cf. also Gillies (Frege, p. 29).

seen, it is argued, from the fact that if one mixes five gallons of water with eight gallons of alcohol, one ends up with less than thirteen gallons of mixture. (Detlefsen would no doubt characterize this phenomenon as another example in which physical reality exhibits arithmetically unstable behaviour.) Let us analyse this reasoning, using our logical system. To simplify matters, let us suppose that we have in room *B* at time *t* five separate one-gallon cans, each filled with water, and eight separate one-gallon cans, each filled with alcohol. And let us suppose the following procedure is followed: firstly, the contents of each of these cans is poured into a large vat; secondly, the mixture is stirred; thirdly, the contents of the vat are poured back into the one-gallon cans. Now evidently it is being thought that what '5 + 8 = 13' asserts is something that implies that thirteen of the one-gallon cans would be filled at the end of the above procedure by the mixture. But let us analyse this reasoning with the above tools. First, we need to add to the interpretation we used earlier

P: x is a one-gallon can in room *B* at time *t* filled with water.

A: x is a one-gallon can in room *B* at time *t* filled with alcohol.

L: x is a one-gallon can in room *B* at time *t* filled with liquid.

We then have as premisses:

[1] $H_5 P$.
[2] $H_8 A$.
[3] $(x)(Lx \longleftrightarrow (Px \lor Ax))$.
[4] $-(\exists x)(Px \ \& \ Ax)$.

And the arithmetical theorem '5 + 8 = 13' can be used to infer

[5] $H_{13} L$.

But this tells us only that there were thirteen gallons of liquid in room *B* at time *t*—something that is not contradicted by experience. Furthermore, if we reinterpret the predicate letters so that

P: x is a one-gallon can of water which was poured into the vat during interval of time i,

A: x is a one-gallon can of alcohol which was poured into the vat during interval of time i,

L: x is a one-gallon can of liquid which was poured into the vat during interval of time i,

we can use the above derivation to infer that thirteen gallons of liquid were poured into the vat during interval of time i. But again this conclusion is not contradicted by experiment. What is contradicted by experiment is the statement that *there will be* thirteen one-gallon cans filled with liquid after the liquid poured into the vat is mixed and poured back into the cans. And this statement cannot be inferred from what is given, just using logic: one needs a statement of physics to the effect that if thirteen gallons of liquid are poured in, thirteen gallons of liquid must result. So the reasonable thing to infer from this example is not the falsity of '5 + 8 = 13' but rather the falsity of the statement of physics that is being presupposed in drawing the conclusion in question.

The above examples illustrate the fundamental way in which a theory about the logical space of open-sentences can take the place of a theory of abstract entities. It can be seen that nothing of practical significance is lost in these cases when, instead of asserting the existence of sets of various kinds, one's logical theory asserts the possibility of constructing open-sentences of various sorts: either way, the usual reasoning involving cardinality that we make use of in ordinary life gets validated. In the next chapter, we shall see how this cardinality theory can be used in ways involving operations and notions other than the straightforwardly logical ones that were discussed in this chapter.

So far, I have said little about cardinality attributes other than the finite ones. What about infinite cardinals? They are simply the cardinality attributes that are not finite cardinality attributes. It is a simple matter to define the appropriate *less than* relation over these attributes and develop the usual type-theoretical version of Cantorian infinite cardinal number theory. Since this can be accomplished without great ingenuity by following what has been done already in standard systems of Simple Type Theory, I leave it as a task for the industrious reader.

6

Measurable Quantities and Analysis

1. A Standard Development of Real Analysis

IT should now be clear, given the finite cardinality attributes and the previously developed method of constructing open-sentences that function as "ordered pairs" of open-sentences, how to carry out in this system the usual construction of open-sentences that function as integers, by taking them to be what are in effect "equivalence classes of ordered pairs of finite cardinality attributes". Similarly, speaking realistically, the rational numbers can be defined to be equivalence classes of ordered pairs of integers; and the real numbers can be taken to be Dedekind cuts of rational numbers. Classical analysis can then be constructed at the levels above the reals. All of this, of course, will require the above sort of transformation of talk about numbers into talk about open-sentences, and there will also be required a significant ascent into higher levels. But the basic principles of development should now be clear.

The above discussion shows how the standard Simple Type theoretical version of analysis and set theory can be reproduced within the constructibility system. Thus, there can be little doubt that any mathematical theory needed in the empirical sciences can be reconstructed within the present system: the many years of mathematical and logical studies following the publication of *Principia Mathematica* have provided us with convincing evidence of this.

However, in order to get clearer on how mathematical representations of physical phenomena get connected to the things we deal with in the physical world, it will be useful to see how certain *measurable quantities*—length and distance, in particular—can be developed within the constructibility framework. A study of measurable quantities naturally brings into view the rational and real number systems. This is not surprising in view of the widely held view that these number systems are used in connection with measurement. But any sort of reasonable measurement theory requires theories about the things to be measured. So in general one would expect applications of the rational and real number systems to involve more than just logical theory. To illustrate both the way in which some measurable quantities can be handled in the present system and also how the open-sentence version of these number systems sketched above can enter into mathematical representations of physical phenomena, I shall present below an axiomatic treatment of lengths.

2. A Theory of Lengths

Henri Lebesgue once advocated a significant change in the French mathematical programme: the sections on fractions, decimals, repeating decimal fractions, and approximation calculations would be replaced by a section on lengths and operations performed on numbers; and this section would then serve as a basis for discussing distances (Measure, p. 35). One reason for dealing with lengths before distances may be that the idea of comparing and measuring lengths is slightly less abstract than the idea of comparing and measuring distances. In any case, I shall follow Lebesgue's suggestion and begin with lengths. This will be a theory about essentially one-dimensional objects (e.g. straight edges of three-dimensional objects or straight lines drawn on paper). The theory is meant to describe idealizations in a relatively naïve, common-sensical way, uncluttered by concerns about quantum mechanical or relativistic phenomena, but sufficiently realistic to throw light on how the rational number system is involved in everyday and scientific concerns. I say 'idealizations' since, strictly speaking, it is questionable that there really are such things as one-dimensional objects—real objects only approximate such idealizations. An explanation and a defence of proceeding in terms of such idealizations will be given in some detail later (after all the axioms of the theory are presented). For now, let us simply note that the theoretical reasoning becomes much simpler by restricting ourselves to these "one-dimensional objects".

The basic principles of the theory will be given as axioms; and although the theory can easily be formalized in the constructibility theory of the present work, I shall present the theory informally, for purposes of perspicuity.

FUNDAMENTALS

The first group of axioms will concern the binary predicates 'G' and 'E', which are to be interpreted intuitively as 'is greater in length than' and 'is equal in length to' respectively. It should be noted, however, that the theory can be intuitively interpreted in many other ways as well (e.g. 'is greater in weight than').

[A1] *Anti-reflexive Law of G*
      Every object x is such that $-xGx$.

[A2] *Transitive Law of G*
      For all objects x, y, and z, if xGy and yGz then xGz.

Comment. The above holds when the theory is given the intuitive interpretation we have in mind. Thus, [A1] holds because no object is longer than itself. The following three axioms assert that E is an equivalence relation—something that is obvious when 'E' is given the intuitive interpretation.

[A3] *Reflexive Law of E*
      Every object x is such that xEx.

[A4] *Symmetric Law of E*
      For all objects x and y, if xEy then yEx.

[A5] *Transitive Law of E*
      For all objects x, y, z, if xEy and yEz, then xEz.

Comment. Of the following three axioms, the first two, the replacement axioms, are self-evident. The third asserts that all the objects in the domain of discourse are comparable with respect to length. That this is so is by no means self-evident. However, just imagine that we have restricted the domain to one-dimensional objects that have length.

[A6] *Replacement*
      For all objects x, y, and z, if xGy and yEz, then xGz.

[A7] *Replacement*
      For all objects x, y, and z, if xGy and xEz, then zGy.

[A8] *Comparability*
      For all objects x and y, either xGy or xEy or yGx.

Comment. If one interprets 'G*' as 'is greater in length than or equal in length to' (defining 'G*' in terms of 'G' and 'E' in the obvious way), one can easily prove that the domain is partially ordered by G*.

The next group of axioms concerns the predicate 'P' which is to be intuitively interpreted as 'is a part of'. Although I provide here only a weak set of "mereological axioms", it is sufficient for my purposes.

[B1] *Reflexive Law of P*
   For every object x, xPx.

[B2] *Antisymmetric Law of P*
   For all objects x and y, if xPy and yPx, then $x = y$.

[B3] *Transitive Law of P*
   For all objects x, y, and z, if xPy and yPz, then xPz.

The symbol 'D' is introduced by a definition. Intuitively, xDy if, and only if, x and y have no parts in common.

Definition. For all objects x and y, xDy if, and only if, it is not the case that there is a z such that zPx and zPy.

When xDy, I shall say that x is *mereologically discrete from y* (or simply: x is discrete from y). The following axiom then says simply that if x is not a part of y, then x must have a part that is discrete from y—something that is clearly true under the interpretation given above.

[B4] For all objects x and y, if $-x$Py, then there is a z such that zPx and zDy.

The following theorems are easily proved:

T1 *Antireflexive Law of D*
   For every object x, $-x$Dx.

T2 *Symmetric Law of D*
   For all objects x and y, if xDy, then yDx.

T3 For all objects x and y, if xDy, then for all objects z and w, if zPx and wPy, then zDw.

Comment. The symbol '$+$' is now introduced by a definition. One can think of $(x + y)$ as the whole that consists of x and y and nothing else discrete from x and y.

Definition. For all objects x, y, and z, $z = (x + y)$ if, and only if, the following holds:
for all w,

   zDw if, and only if, (xDw and yDw)

$(x + y)$ will be called the "mereological sum of x and y". The next group of axioms concerns this operation.

Comment. The reader should note that the mereological sum of x and y should not be taken to be "the putting together of x and y". Thus, if x and y are

———————————   ————

respectively, then the mereological sum of x and y is that "object" consisting of those two line segments separated by the above space.

[C1] For all objects x, y, z, and w, if xEy, zEw, xDz, and yDw, then
$(x + z)E(y + w)$.

[C2] For all objects x and y, xGy if, and only if, there is a z and a w such that zDw and $x = (z + w)$ and zEy.

Comment. [C1] asserts that equals added to equals yields equals, so long as the things being added are discrete from the things to which they are being added. (See figure below.) One can think of [C1] as telling us that we determine the length of $(x + z)$ by, in effect, putting z end to end with x in the natural fashion:

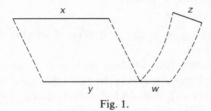

Fig. 1.

The 'only if' part of [C2] reflects the intuitively compelling idea that if y is smaller in length than x, then y must be equal in length to some proper part of x. The 'if' part of [C2] informs us that every part of a whole contributes something to the whole's length, and it reflects the intended interpretation of the system according to which there are no points which are parts of a whole. To see this, suppose that there are points in the domain and imagine that we have two lines x and y which are such that x differs from y only by virtue of the fact that x includes an end-point that is not a part of y. One might think of x as the closed interval $[0, 1]$ and of y as the half-open interval $[0, 1)$. In this case, there would be a z (that is, y) and a w (that is, the end point of x) which are such that zDw, $x = z + w$, and zEy. But x would not be greater in length than y. Thus, [C1] tells us something about how to determine the lengths of mereological sums, and [C2] requires the domain to contain only things with lengths (no "points"). [C1] is true in virtue of the notion of length being articulated; [C2] is true because *that* is how we interpret the system: we interpret the theory in such a way that points are not regarded as parts.

It needs to be emphasized that no claim is being made here that the notions of parts, sums, etc. being articulated are the notions of ordinary people or are those best suited to scientific work. In particular, it is certainly *not* being claimed that the mathematician's way of regarding lines as sets of points should be rejected on the grounds that it doesn't correspond to our "ordinary notion" of lines.

I turn now to some theorems. The proofs of the first three of the following, I leave as simple exercises for the reader.

T4 For every object x, $(x+x)=x$.

T5 *Commutative Law of* +
For all objects x and y, $(x+y)=(y+x)$.

T6 *Associative Law of* +
For all objects x, y, and z,
$x+(y+z)=(x+y)+z$.

T7 For all objects x, y, z, and w, if xGy, zEw, xDz and yDw, then $(x+z)G(y+w)$.
To demonstrate this, assume that x, y, z, and w satisfy the antecedent. Then, since xGy, by [C2], for some objects s and t, we have $x=(s+t)$, sDt, and sEy. Now since xDz, then by T3, sDz. So by [C1], $(s+t)E(y+w)$. Then since
$(x+z)=((s+t)+z)$
$(x+z)=((t+s)+z)$ by T5
so it $=(t+(s+z))$ by T6.
Since tDz and tDx, by T3, we have $tD(s+z)$. Hence, by [C2], we have $(x+z)G(y+w)$.

T8 For all objects x, y, z, and w, if xGy, zGw, xDz, and yDw, then $(x+z)G(y+w)$.

This theorem follows straightforwardly from T7.

In the following, a new predicate 'EP' is introduced; it can be thought of as relating "properties" (in the sense of the previous chapter) to objects. It is to be intuitively interpreted as meaning 'is an equi-partition of', which I shall explain below.

Definition. For every property F and every object x, F is a *partition* of x if, and only if, the following hold:

1. F has finite cardinality (say n);
2. for all objects y, if y satisfies F then yPx;
3. for all objects y and z, if both y and z satisfy F, then either yDz or $y = z$;
4. the sum $(x_1 + x_2 + \ldots + x_n) = x$, where the x_i are the n parts of x that satisfy F.

Comment. Roughly speaking, a partition of x breaks x into n parts that are pairwise discrete. An equi-partition of x (defined below) is a partition of x that yields parts that are the same length.

Definition. For every property F and every object x, F is an equi-partition of x if, and only if, F is a partition of x and, in addition, all the parts of x that satisfy F are equal in length, i.e. add to the above four conditions:

5. for all objects y and z, if both y and z satisfy F, then yEz.

The last axiom of this theory, the *Equi-Partition Axiom*, can now be given:

[D1] For every object x and for every finite cardinality attribute **N**, there is a property F such that FEPx and F has cardinality attribute **N**.

Discussion. [D1] is a strong axiom and tells us that the objects in the domain are all "continuous" (or potentially infinitely divisible in principle into equal parts). If my purposes were different, I would probably use different axioms in place of [D1], and then prove [D1] as a theorem. However, since taking [D1] as an axiom allows me to avoid some unwieldy derivations and to go almost directly into the material involving the "rationals", I shall forgo such axiomatic explorations here.

It may be asked, however, whether this axiom is true. How do we know, in particular, that each object has infinitely many parts of arbitrarily small size? Might we not find, for example, from scientific investigations, that there is a limit to how small the parts of an object can be? In response, it needs to be emphasized that it is not being maintained here that the axioms of this theory are, in fact, true of genuine physical objects. This theory provides the conceptual framework for speaking of relations of length between all possible objects of finite length, from the microscopically small to the astronomically enormous. Thus, the present theory attempts to capture a way of thinking about lengths that is close to our naïve views about lengths and that provides a basic framework for reasoning about lengths in a coherent way. So we can regard the Equi-Partition Axiom as an expression of how we are to view the objects we talk about in this theory, whether or not it makes sense to regard actual physical objects as actually having infinitely many parts. It is for purposes of conceptual simplicity that we shall regard objects (and analyse objects) as infinitely divisible into parts of arbitrarily small size.

But am I not, in effect, committing myself to the existence of strange entities that differ only slightly from the abstract mathematical objects of the mathematical Platonist? In fact can we not regard mereological sums of one-dimensional objects as abstract entities? After all, we cannot expect to find such things anywhere in space.

One important respect in which my position, *vis-à-vis* the "objects" talked about in my theory of lengths, differs from the mathematical Platonist's, *vis-à-vis* the "objects" talked about in set theory, is this: I do not commit myself to the existence of the mereological sums of one-dimensional objects, since I do not assert the truth of the axioms of the theory. In this connection, let us review some aspects of Quine's doctrine of "ontological commitment". According to Quine, one does not commit oneself to a certain sort of entity F just by considering a theory that is committed to Fs. Nor does one commit oneself to Fs by developing a theory that commits itself to Fs. It is only by asserting or maintaining the truth of such a theory that one commits oneself. But in the above case, I do not assert that the axioms are true. I do not claim that there really are such one-dimensional objects. So I do not commit myself to their existence.

To appreciate the position I am taking, it should be recalled that it is not at all unusual for scientists to develop theories by making *idealizations* of actual situations. Consider what one philosopher of science has claimed recently about these idealizations:

Most philosophers have made their peace with idealizations; after all, we have been using them in mathematical physics for well over two thousand years. Aristotle in proving that the rainbow is no greater than a semi-circle in *Meteorologica* III. 5 not only treats the sun as a point, but in a blatant falsehood puts the sun and the reflecting medium (and hence the rainbow itself) the same distance from the observer. Today we still make the same kinds of idealizations in our celestial theories. Nevertheless, we have managed to discover the planet Neptune, and to keep our satellites in space. Idealizations are no threat to the progress of science (Cartwright (Lie), pp. 110–11).

Furthermore, these idealizations may involve apparent reference to various fictional *idealized* entities: the physicist may talk about completely rigid bodies, absolutely incompressible fluids, frictionless planes, etc. Consider, for example, the following discussion by the physicists Robert Lindsay and Henry Margenau of the notion of a *particle* (in their (F of P), p. 80):

A *particle* is an entity assumed to possess several properties the most fundamental of which are as follows:
1. A particle has *position* . . . but no extension; . . .
2. A particle has *inertia*, the measure of which is *mass*. . . .
3. A particle will have certain relations to other particles—gravitation, etc.
It must be remarked with respect to the above that the particle is a *symbolic* conception like all conceptions in physics. It is, of course, suggested by our experience but represents an idealization of that experience.

Now do these distinguished physicists believe in strange extensionless entities called particles? Not at all. They may develop a theory that talks about particles, but the theory is not intended to be a literal description of what truly exists. The particle is an *idealization*. When they apply their theory, they will take into account the fact that there are no genuine particles, but that certain real things do, in the relevant respects, approximate particles—these may have properties that are similar to those attributed to particles and the respects in which these real physical entities differ from particles may be negligible in certain sorts of contexts. In this respect, the physicists' theory of particles is different from the set-theoretical Platonist's views about sets. The latter does not treat the axioms of set theory as literally false. Nor does this theoretician believe that the value of the set theory derives from the fact that certain real objects approximate sets. (But cf. Kitcher's view of set theory discussed in Chapter 11, Section 2.)

But why construct a theory of idealized objects? Why not develop a theory of actual physical objects instead? The mathematician-scientist Henri Poincaré has responded to such questions (in (*Value*), p. 125). Imagine, he suggests, that we are attempting to describe the relationships that obtain between two physical objects A and B. We find, however, that these relationships are extremely complex. So instead of attempting to describe these relationships directly, we describe the relationships that certain fictional ideal objects A' and B' have to one another instead. The idea is to select the fictional objects in such a way that the relationships that obtain between A' and B' are relatively simple and easily described. But we also require that the differences between A and A', on the one hand, and B and B', on the other, are relatively small. Thus, the complicated relationships are "replaced" by relationships that are relatively simple.

For example, if A and B are two natural solid bodies which are displaced with slight deformation, we envisage two movable rigid figures A' and B'. The laws of the relative displacements of these figures A' and B' will be very simple; they will be those of geometry. And we shall afterwards add that the body A, which always differs very little from A', dilates from the effect of heat and bends from the effect of elasticity. These dilations and flexions, just because they are very small, will be for our mind relatively easy to study. Just imagine to what complexities of language it would have been necessary to be resigned if we had wished to comprehend in the same enunciation the displacement of the solid, its dilation, and its flexure? (*Value*, p. 125).

Let us reconsider, in the light of the above discussion, the theory of lengths being developed here. We could have developed a theory of lengths of actual physical objects instead of idealized one-dimensional objects. But this would have involved considerable complications. Imagine, for example, that we are attempting to analyse the statement that Zubin Mehta's baton is twice as long as the standard metre in Paris. As a first approximation, we might analyse the statement as meaning that we can divide the baton into two non-overlapping parts each of which is equally as long as the standard metre in Paris. But this clearly won't do, since if that is what the statement means, then it would seem that the baton would also be *four times as long* as the standard metre. For we could just as well divide the baton into four non-overlapping parts, each of which is equally as long as the standard metre. We could do this by dividing each of the two original parts into two parts, making the division along the axis of the baton thus:

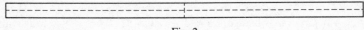

Fig. 2.

To avoid such absurdities, we would have to make some such further specification as that the "parts" we talk about are to be cross-sections perpendicular to the axis of the baton. But how are we to specify such a condition in general, i.e. for any object to which we want to attribute length? In particular, how are we to specify the axis to which the cross-sections are to be perpendicular? Any relatively precise and general answer will be complicated. One possible way of handling this problem is to regard 'equally long' as denoting a four-ary relation: an object x relative to axis y is equally long as object z relative to axis w. But obviously such an analysis would involve us in many complications.

This is just one of many difficulties of a similar kind that must be dealt with if one is to develop a theory of lengths in terms only of real physical objects. These difficulties, however, are by no means insuperable. They are just messy and tedious. And adding a number of complicated and elaborate definitions to deal with such problems will obscure the central ideas of length that one wishes to clarify. How much easier and simpler it is to deal with (imaginary) one-dimensional objects.

But how do the real physical objects to which we wish to apply length predicates differ from the one-dimensional ideal objects of the theory? Real physical objects will not have sharp boundaries, they will not be one-dimensional, they may be curved and bent in various ways. But we can distinguish among physical objects some that are better approximations to one-dimensional objects than others. For example, a line drawn on paper using straight edge and pencil is a better approximation to a (one-dimensional) straight line segment than a football. A conductor's baton is a better approximation than an oak tree. Furthermore, we can see in what respects these approximations are not truly one-dimensional or are not truly straight. These facts are important in seeing how the theory in question is to be applied. For it will be obvious that the sort of "parts" we distinguished in discussing the baton example is not what we had in mind in providing the above analysis of the statement that Zubin Mehta's baton is twice as long as the standard metre. After all, we simply cannot distinguish four such "parts" when dealing with one-dimensional objects.

Consider another sort of problem that arises when one is dealing with real physical objects instead of the idealized ones. According to the rough intuitive analysis provided earlier, to say that the baton is twice as long as the standard metre is to say that the baton is divisible into two non-overlapping parts . . . But what does 'divisible' mean here? Suppose, for example, that there are certain physical objects that are physically *indivisible* into parts, i.e. these objects cannot be physically cut up into parts. Still, could not such an object be twice as long as the standard metre in Paris? Surely, all that would be required is that one be able to *distinguish* these non-overlapping parts, say by placing an accurate metre stick next to the baton and marking off the two parts. Besides, we will wish to talk about the lengths of phenomena that cannot reasonably be regarded as physical objects at all, such as, say vapour trails left by a jet airplane or a beam of light produced by a flashlight in a darkened room; and there may not be any clear sense in which we can "physically divide such phenomena into parts". Indeed, it may not even be clear what it would mean to say that such phenomena have "parts" that are non-overlapping and equally long. Again, these problems are not insuperable, but attempting to deal with them in sufficient generality to make the theory enlightening and useful would require making complicated definitions and conditions, all of which would tend to obscure the central ideas of length that one is attempting to clarify.

I wish to show that we do not have to believe in the existence of lengths in order to be able to talk about lengths and espouse theories in which lengths are talked about in pretty much the ordinary way we do. The idea is to show how talk about the length of an object can be analysed as a manner of speaking about the objects that are said to have length. My strategy has been to sketch the main ideas of this analysis by dealing with idealized fictional objects—this in the name of simplicity and clarity. However, since many real objects approximate the fictional objects of the theory, the reader can see, at least in a general way, how the analysis can be carried over to the case of ordinary physical objects.

> T9 For all properties F and H, and all objects x, if $F\mathbf{EP}x$ and $H\mathbf{EP}x$ and if the following holds
>
> > for all objects y and z, if Fy and Hz then $y\mathbf{E}z$
>
> then for some finite cardinality attribute $\mathbf{N}$, F and H both have $\mathbf{N}$.

Assume that F has cardinality $\mathbf{M}$, which is greater than the cardinality $\mathbf{N}$ of H. Then we have, as parts of x that satisfy F

$$x_1, x_2, \ldots, x_m$$

and as parts of x that satisfy H

$$y_1, y_2, \ldots, y_n$$

But $x_1\mathbf{E}y_1$ and $x_2\mathbf{E}y_2$. Furthermore, $x_1\mathbf{D}x_2$ and $y_1\mathbf{D}y_2$. Hence by [C1], $(x_1 + x_2)\mathbf{E}(y_1 + y_2)$.

Continuing in this way, we obtain

$$(x_1 + x_2 + \cdots + x_m)E(y_1 + y_2 + \cdots + y_m)$$

Hence,

$$xE(y_1 + y_2 + \cdots + y_m)$$

But we also have $xE(y_1 + \cdots + y_n)$. And since, $xE((y_1 + \cdots + y_m) + (y_{m+1} + \cdots + y_n))$, we can conclude from [C2], xGx. This contradicts T1.

T10 For all properties F and H, and every object x, if $FEPx$ and $HEPx$ and there is an attribute of finite cardinality N which is such that both F and H have N, then for all objects y and z, if y satisfies F and z satisfies H, then yEz.

To see that this is so, assume that F, H, and x satisfy the antecedent; and assume, for a *reductio* argument, that there are objects y and z which are such that y satisfies F and z satisfies H, but that it is not the case that yEz. We can assume, without loss of generality, that yGz. Let $x_1, \ldots, x_n$ be the parts of x that satisfy F and let $y_1, \ldots, y_n$ be the parts of x that satisfy H. Then using T7, it can be shown that $x_1Gy_1, \ldots, x_nGy_n$. Then, using T8, we obtain

$$(x_1 + \cdots + x_n)G(y_1 + \cdots + y_n)$$

Then xGx, which contradicts T1.

T11 For all properties F and H, and all objects x and y, if xEy, $FEPx$ and $HEPy$, and there is a finite cardinality attribute N which is such that both F and H have N, then for all objects z and w, if z satisfies F and w satisfies H, then zEw.

To show this, one need only proceed as above with only minor differences.

RATIONAL LENGTHS

I now introduce a four-ary predicate 'E' that is intended to hold of objects x and y and finite cardinality attributes N and M when x is n/m as long as y.

Definition. For all objects x and y, and for all finite cardinality attributes N and M, $xE[N/M]y$ if, and only if, there are properties F

and H which are such that $FEPx$ and $HEPy$ and F has $\mathbf{N}$ and H has $\mathbf{M}$ and for all objects y and z, if Fy and Hz, then yEz.

(*Note.* For purposes of perspicuity, I shall state the following theorems using constants instead of variables. It should be understood, however, that the appropriate generalizations of the statements are intended as the theorems. In addition, I shall only give the main ideas of the "proofs", using for the most part a minimum of notation.)

T12 If aEb and $bE[\mathbf{M}/\mathbf{N}]c$, then $aE[\mathbf{M}/\mathbf{N}]c$.

Assume the antecedent. Then there are equi-partitions F and H of b and c respectively, such that F has cardinality $\mathbf{M}$, H has cardinality $\mathbf{N}$, and all the parts that satisfy the partition are of equal length. By [D1], there is an equi-partition J of a which is such that J has cardinality $\mathbf{M}$. By T11, each part of a that satisfies J has the same length as each part of c that satisfies H. Hence $aE[\mathbf{M}/\mathbf{N}]c$.

T13 If $aE[\mathbf{M}/\mathbf{N}]b$, then $aE[\mathbf{M} \times \mathbf{P}/\mathbf{N} \times \mathbf{P}]b$.

Assume the antecedent. Then there are equi-partitions F and H of a and b respectively such that F has cardinality $\mathbf{M}$, H has cardinality $\mathbf{N}$, and all the parts that satisfy either F or H are of equal length. But by [D1], there are equi-partitions F_1, F_2, ..., F_m of each part of a that satisfies F and which are such that each F_i has cardinality $\mathbf{P}$. Similarly, there are equi-partitions H_1, H_2, ..., H_n of each part of b that satisfies H and which are such that each H_i has cardinality $\mathbf{P}$. It follows that there is an equi-partition F^* of a which has cardinality $\mathbf{M} \times \mathbf{P}$ and an equi-partition H^* of b which has cardinality $\mathbf{N} \times \mathbf{P}$ and which are such that all the parts of a that satisfy F^* have the same length as every part of b that satisfies H^*. Hence $aE[\mathbf{M} \times \mathbf{P}/\mathbf{N} \times \mathbf{P}]b$.

T14 If $aE[\mathbf{M} \times \mathbf{P}/\mathbf{N} \times \mathbf{P}]b$, then $aE[\mathbf{M}/\mathbf{N}]b$.

Assume the antecedent. Then there are equi-partitions F and H of a and b respectively such that F has cardinality $\mathbf{M} \times \mathbf{P}$ and H has cardinality $\mathbf{N} \times \mathbf{P}$ and the parts of a that satisfy F have the same length as the parts of b that satisfy H. Let $a_1, a_2, \ldots, a_{mp}$ be the parts of a that satisfy F. Then let

$$a^*_1 = a_1 + a_2 + \cdots + a_p$$
$$a^*_2 = a_{p+1} + a_{p+2} + \cdots + a_{2p}$$

.
.
.

$$a^*_m = a_{(m-1)p+1} + \ldots + a_{mp}$$

and let J be a property satisfied by all and only the a^*_i. It is a simple matter to prove that J is an equi-partition of a. Now the analogous reasoning in the case of b yields an equi-partition K which has cardinality $\mathbf{N}$. One needs only to show that each part of a that satisfies J has the same length as each part of b that satisfies K.

T15 $a\mathrm{E}[\mathbf{M}/\mathbf{N}]b$ if, and only if, $a\mathrm{E}[\mathbf{M} \times \mathbf{P}/\mathbf{N} \times \mathbf{P}]b$.

T16 If $a\mathrm{E}[\mathbf{M}/\mathbf{N}]b$ and $a\mathrm{E}[\mathbf{P}/\mathbf{Q}]b$, then $\mathbf{M} \times \mathbf{Q} = \mathbf{N} \times \mathbf{P}$.

Assume the antecedent. Then by T15, $a\mathrm{E}[\mathbf{M} \times \mathbf{Q}/\mathbf{N} \times \mathbf{Q}]b$ and $a\mathrm{E}[\mathbf{N} \times \mathbf{P}/\mathbf{N} \times \mathbf{Q}]b$. So there are equi-partitions F and H of a and b respectively such that F has cardinality $\mathbf{M} \times \mathbf{Q}$ and H has cardinality $\mathbf{N} \times \mathbf{Q}$. Similarly, there are equi-partitions J and K of a and b respectively such that J has cardinality $\mathbf{N} \times \mathbf{P}$ and K has cardinality $\mathbf{N} \times \mathbf{Q}$. Since H and K have the same cardinality, it follows from T10 that the parts of b that satisfy H have the same length as the parts of b that satisfy K. Hence the parts of a that satisfy F have the same length as the parts of a that satisfy J. Hence from T9, F and J have the same cardinality. Thus, $\mathbf{M} \times \mathbf{Q} = \mathbf{N} \times \mathbf{P}$.

T17 If $a\mathrm{E}[\mathbf{M}/\mathbf{N}]b$ and $\mathbf{M} \times \mathbf{Q} = \mathbf{N} \times \mathbf{P}$, then $a\mathrm{E}[\mathbf{P}/\mathbf{Q}]b$.

Assume the antecedent. By T13, we have $a\mathrm{E}[\mathbf{M} \times \mathbf{Q}/\mathbf{N} \times \mathbf{Q}]b$. Then $a\mathrm{E}[\mathbf{N} \times \mathbf{P}/\mathbf{N} \times \mathbf{Q}]b$. Then by T14, we have $a\mathrm{E}[\mathbf{P}/\mathbf{Q}]b$.

T18 If $a\mathrm{E}[\mathbf{M}/\mathbf{N}]b$, then $a\mathrm{E}[\mathbf{P}/\mathbf{Q}]b$ if, and only if, $\mathbf{M} \times \mathbf{Q} = \mathbf{N} \times \mathbf{P}$.

T19 If $a\mathrm{E}[\mathbf{M}/\mathbf{N}]b$ and $c\mathrm{E}[\mathbf{P}/\mathbf{Q}]b$, then $a\mathrm{G}c$ if, and only if, $\mathbf{M} \times \mathbf{Q} > \mathbf{N} \times \mathbf{P}$.

The justification of T19 is straightforward and left as an exercise. T18 and T19 should bring to mind the criterion of identity of rational numbers

$$m/n = p/q \text{ if, and only if, } mq = np$$

and the rule

$$m/n > p/q \text{ if, and only if, } mq > np.$$

When the properties of rational numbers are viewed as being

abstracted from those of "rational lengths" (to be explained below), certain features of the rational number system that appear to be purely arbitrary at first, can be seen to be necessary.

To sketch the main ideas of how we obtain a theory of *rational lengths* (which are analogues of rational numbers), I shall use realistic terminology for the sake of simplicity and perspicuity, although it should be clear by now that this whole discussion could be carried out by speaking of the constructibility of open-sentence tokens instead of the existence of classes. Now we know from [A3]–[A5] that 'E' expresses an equivalence relation over the domain of objects. So we can partition the domain of objects into mutually exclusive and exhaustive equivalence classes, just as, in the case of cardinality, we partitioned the domain of properties into equivalence classes using the equivalence relation of equinumerosity. So if we let $[x]$ be the equivalence class of all objects that have the same length as x, we can treat $[x]$ as *the length of* x, just as the equivalence class of all properties equi-numerous with F can be treated as the cardinality of F. Then if we order the lengths according to the *intuitively natural rule*

$$[a]\mathbf{G}[b] \text{ if, and only if, } a\mathbf{G}b.$$

$\mathbf{G}$ can be easily shown to be a linear ordering of the lengths. Now suppose we select some object u to be the standard to which all other objects will be compared with respect to length (somewhat analogously to the way the standard metre in Paris is used), and suppose that we lay down a hypothesis that will function in this theory in a way analogous to the way the Hypothesis of Infinity functions in the theory of finite cardinality. In particular, suppose we take as a hypothesis:

For all finite cardinality attributes $\mathbf{M}$ and $\mathbf{N}$, there is an object x such that $x\mathbf{E}[\mathbf{M}/\mathbf{N}]u$.

The idea here is that we shall be concerned henceforth with a domain of objects in which there are objects of all possible (rational) lengths. Thus we will have all the possible rational lengths, and these lengths will be linearly ordered when we use the relation $\mathbf{G}$ given by what we called above "the intuitively natural rule".

Let us use the notation '$[m/n]$' to stand for the equivalence class of all objects x such that $x\mathbf{E}[\mathbf{M}/\mathbf{N}]u$. Then x is a rational length if, and only if, x is one of these equivalence classes. T19 tells us that in this natural ordering, $(m/n)\mathbf{G}[p/q]$ if, and only if, $mq > np$, which is essentially the rule we use to order the rationals. Similarly, T18 tells us that $[m/n] = [p/q]$ if, and only if, $mq = np$, which is essentially the criterion of identity of rationals mentioned above. In short, from the point of view of the theory of lengths sketched above, the rule of ordering rationals and the criterion of identity of rationals are not arbitrary, but instead reflect necessities imposed on us by the concepts we work with.

With the above notational conventions, we can state as a simple corollary to T19:

T19 If aE[**M/N**]u and bE[**P/Q**]u, then aGb if, and only if, [m/n]G[p/q].

T20 If aE[**M/N**]b, cE[**P/N**]b, and aDc, then $(a+c)$ E[(**M** + **P**)/**N**]b.

Assume the antecedent. Then there are equi-partitions F and H of a and b respectively such that F has cardinality **M** and H has cardinality **N**. Similarly there are equi-partitions J and K of c and b respectively such that J has cardinality **P** and K has cardinality **N**. Let O be an open-sentence satisfied by all and only those objects that satisfy F or K or both. O can be straightforwardly shown to be an equi-partition of $(a+c)$ and to have cardinality **M** + **P**.

T21 If aE[**M/N**]b, cE[**P/Q**]b, and aDc, then $(a+c)$ E[(**M** × **Q** + **N** × **P**)/**N** × **Q**]b.

Assume the antecedent. Then by T15, aE[**M** × **Q**/**N** × **Q**]b. Similarly, cE[**N** × **P**/**N** × **Q**]b. So by T19, we obtain the consequent.

Addition of Lengths

What would be a reasonable way of regarding the addition of lengths? Well, consider the addition of two cardinals **M** and **N**. Recall that we took representative properties that have the cardinals, added them together (i.e. took the union) and then took the cardinality of the result to be the arithmetical sum. It would be natural to do the analogous thing in the case of the addition of lengths: we take representative objects that have the lengths in question, add these objects together and get the lengths of the result. But what should we count as "adding the objects together"? The natural operation here would be to take the *mereological sum* of the objects. And just as, in the case of cardinals, we needed to require the representative properties to be disjoint, so also in this case we need to require the representative objects to be mereologically discrete from one another. When we take the addition of rational lengths in this way, T21 shows us that we get essentially the standard rule for the addition of rationals, i.e. we get:

$$[m/n] + [p/q] = [(mq + np)/nq].$$

Multiplication of Lengths

Consider a simple case of multiplication of lengths. We have some object a that is four units long. We want something that is three times as long as a. If b is such an object, how many units long would it be? Clearly, we find the answer by multiplying 3×4. And in general, we use multiplication to find out how long an object is that is m times as long as something that is n times as long as u. Now consider the more general case in which an object is m/n times as long as something that is p/q times as long as u. How long would the object be, relative to u? The answer is given by the following theorem.

T22 If $a\mathbf{E}[\mathbf{M}/\mathbf{N}]b$ and $b\mathbf{E}[\mathbf{P}/\mathbf{Q}]c$, then $a\mathbf{E}[\mathbf{M} \times \mathbf{P}/\mathbf{N} \times \mathbf{Q}]c$.

I leave this as a simple exercise.

REAL LENGTHS

There are well-known reasons for expanding the system of rational lengths. Euclid's geometry can be used to uncover one such reason. A right triangle with two sides each of length $[1/1]$ cannot, in this geometry, have a hypotenuse which is a rational length, i.e. the hypotenuse cannot have a length $[m/n]$. To fill in the gaps, we can define real lengths to be Dedekind cuts of rational lengths. I briefly sketch below the main ideas. (*Note.* In the following, even though the real-length open-sentences are of a level higher than that of rational lengths, this difference in levels between the rational and real lengths is not marked by the style of variable employed.)

Definition. R is a real-length open-sentence if, and only if,

(*a*) for all rational-length open-sentences F and H, if F satisfies R and F is greater than H (by the ordering relation $\mathbf{G}$), then H satisfies R;

(*b*) it is not possible to construct a rational-length open-sentence F which both satisfies R and is such that all rational-length open-sentences that satisfy R are less than or equal to F;

(*c*) it is possible to construct a rational-length open-sentence that does not satisfy *R* and also one that does satisfy *R*.

The totality of real-length open-sentences is automatically sorted into equivalence classes by extensional identity (i.e. the I-predicate) to give us the real lengths. We can now order the real lengths by the rule

FG^*H if, and only if, all rational lengths that satisfy *H* satisfy *F*, but not all rational lengths that satisfy *F* satisfy *H*.

It is easy to show that the real lengths are linearly ordered by this relation. A real length *F* is an *upper bound* of a set *S* (i.e. an appropriate open-sentence) of real lengths if, and only if, every member of *S* is less than *F*. A real length *F* is a *least upper bound* of a set *S* of real lengths if, and only if, every upper bound of *S* is greater than or extensionally identical to *F*. The *completeness* of the real lengths can be proved in the standard way, i.e. it can be proved that every non-empty set of real lengths that has an upper bound is a least upper bound.

The *Sum of Real Lengths

If *a* and *b* are real lengths, $a + {}^*b$ is an open-sentence *F* satisfying the following condition:

x satisfies *F* if, and only if, it is possible to construct rational lengths *r* and *s* such that *r* satisfies *a*, *s* satisfies *b*, and *x* is extensionally identical to $r + s$.

$a + {}^*b$ will be called a *-sum of *a* and *b*.

The following can then be shown:

1. Any two *-sums of real lengths *a* and *b* are extensionally identical.

2. For any real lengths *a* and *b*, it is possible to construct a *-sum $a + {}^*b$.

3. Every *-sum of real lengths is a real length.

4. The operation of *-addition of real lengths is associative and commutative.

(I do not give proofs, since with only minor adjustments, they can be constructed from proofs in standard mathematics text-books.)

The Product of Real Lengths

If *a* and *b* are real lengths, then $a \times {}^*b$ is an open-sentence *F* satisfying the following condition:

x satisfies F if, and only if, it is possible to construct rational lengths r and s such that r satisfies a, s satisfies b, and x is extensionally identical to $r \times s$.

$a \times {}^*b$ will be called a *-product of a and b.

The following can be shown:

1. Any two *-products of a and b are extensionally identical.
2. For any real lengths a and b, it is possible to construct a *-product $a \times {}^*b$.
3. Every *-product of real lengths is a real length.
4. *-multiplication of real lengths is associative, commutative, and distributive over *-addition.

Images of Rational Lengths

If r is a rational length, then let r^* be a real length that is satisfied by all and only rational lengths that are less than r. r^* will be said to be an *image* of r. The following shows that the system of rational lengths is isomorphic to the system of images of rational lengths.

T23   For all rational lengths a and b, and for all $(a + b)^*$, a^* and b^*,
$$(a+b)^* = a^* + b^*$$
$$(a \times b)^* = a^* \times b^*$$
$b\mathbf{G}a$ if, and only if, $b^*\mathbf{G}^*a^*$

where '$=$' signifies extensional identity.

As simple exercise, the reader can prove:

T24   *Density*: If a and b are real lengths, then $b\mathbf{G}^*a$ if, and only if, there is an image c of a rational length which is such that $b\mathbf{G}^*c$ and $c\mathbf{G}^*a$.

Now how are these "real lengths" I have been talking about related to the "objects" in the domain that supposedly have lengths? The following discussion will bring out the connection.

Definition. An object x *has real length* r if, and only if, x is greater in length than every object that has a rational length that satisfies r and x is not greater in length than any object that has a rational length that does not satisfy r.

Only the last of the next three theorems will be proved.

T25a   If objects a and b have real lengths x and y respectively, then $y\mathbf{G}^*x$ if, and only if, b is greater in length than every object that has a rational length satisfying x but a is not greater in length than every object having a rational length satisfying y, where the relation *is greater in length than* is the relation G axiomatized earlier.

T25b If objects a and b have real length r, then $a\mathbf{E}b$.

T25c If a is a rational length and b is a real length, then $b\mathbf{G}^*a^*$ if, and only if, a satisfies b.

I first show that the left-to-right conditional holds. Start by assuming that $b\mathbf{G}^*a^*$. Then every rational length that satisfies a^* satisfies b. Now suppose, for a *reductio* argument, that a does not satisfy b. Then, from the definition of real length, there could be no rational length c which is such that $c\mathbf{G}^*a^*$ and c satisfies b^*. Hence, $b^* = a^*$. Impossible. To show that the right-to-left conditional holds, suppose that a satisfies b. Then, for all rational lengths c which are such that $a\mathbf{G}c$, c satisfies b. Hence, $b\mathbf{G}^*a^*$ or $b = a^*$. If $b = a^*$, then a satisfies a^*. But that is impossible.

I now state and demonstrate the main theorem.

T26 If object a has real length x and object b has real length y, then $b\mathbf{G}a$ if, and only if, $y\mathbf{G}^*x$.

To show that this is so, I consider the four possible cases:

1. *x and y are images of rational lengths s and t respectively.* Then by T19a, $b\mathbf{G}a$ if, and only if, $t\mathbf{G}s$. By T23, $t\mathbf{G}s$ if, and only if, $t^*\mathbf{G}^*s^*$. So $b\mathbf{G}a$ if, and only if, $y\mathbf{G}^*x$.

2. *x is an image of some rational length s: y is not an image*

Suppose $y\mathbf{G}^*x$. By T25c, s satisfies y. Then by the definition of 'has real length', $b\mathbf{G}a$.

Suppose $b\mathbf{G}a$. Then s must satisfy y, for otherwise b would be greater in length than something, namely a, that has a rational length not satisfying y. Then every rational length that is less than s must also satisfy y (from the definition of real lengths). Hence $-x\mathbf{G}^*y$, i.e. either $x = y$ or $y\mathbf{G}^*x$. But $x = y$ is impossible.

3. *y is an image of some rational length t; x is not an image*

Suppose that $y\mathbf{G}^*x$. Then by Comparability either $b\mathbf{G}a$ or $b\mathbf{E}a$ or $a\mathbf{G}b$. But b can't be equal in length to a, because of the conditions we are assuming that x and y satisfy. If $a\mathbf{G}b$, then by case (2) above, $x\mathbf{G}^*y$. But that contradicts the above supposition. Hence, $b\mathbf{G}a$.

Suppose $-y\mathbf{G}^*x$. Then either y is extensionally identical to x or $x\mathbf{G}^*y$. The first alternative is impossible. So consider the second. Then, by case (2), we have $a\mathbf{G}b$, and hence $-b\mathbf{G}a$.

4. *Neither x nor y is an image*

Suppose yG^*x. Then by density, there is a rational length s such that yG^*s^* and s^*Gx. Let c be an object having rational length s. Then bGc and cGa (by the previous two cases). Thus, bGa.

Suppose bGa and $-yG^*x$. Then either x is extensionally identical to y or xG^*y. If xG^*y, then from the first part of this case, we have aGb, which contradicts bGa. If y is extensionally identical to x, then every rational length satisfying y satisfies x, and conversely. So b is longer than every object having a rational length satisfying x, and b is not longer than any object having a rational length not satisfying x. Hence b has the length of x, and, by T25b, bEa. This contradicts bGa.

3. Geometrical Representations of Functions

Distance is obviously similar to length. Indeed, a theory of distance can be developed within the framework of Euclid's geometry described earlier, essentially by adding to the geometrical axioms the theory of lengths articulated in this chapter: the line segments discussed in Euclid's constructive theory, i.e. the line segments it is possible to construct, would serve to provide us with the "objects" in terms of which a distance function could be defined, for the distance between two points a and b can be taken to be the length of a line segment connecting a with b. However, since I shall be concerned, in this section, with correlating points on a plane with ordered pairs of real numbers, we need not concern ourselves with distances: we can set up the correlation directly in terms of lengths.

Let us begin with some line A in a plane and a point o in A. We can construct, in that plane, a line B that is perpendicular to A and that intersects o. Take A to be the x-axis and B to be the y-axis. The x-axis is divided by o into two segments, one of which we can take to be the positive part of the x-axis, and the other of which we can take to be the negative part. I shall call those line segments that are parts of the positive part of the x-axis and that have o as one end-point '*finite positive segments*'. Let u be some arbitrarily chosen finite positive segment. Then I shall call those finite positive segments that have rational length (relative to u) '*rational segments*'.

The rational segments are correlated one-one to rational numbers by an appropriate open-sentence $c_1(x)$ which takes a given rational segment s to the rational number r when, and only when, s and r are related by the condition:

$s\mathbf{E}[\mathbf{M/N}]u$ and $r = \mathbf{M/N}$.

Then an appropriate open-sentence $c_2(x)$ will correlate one-one the finite positive segments with the positive real numbers by taking any finite positive segment s to the positive real number r when and only when s and r satisfy the condition:

for every rational segment z, $s\mathbf{G}z$ iff $c_1(z)\epsilon r$.

The open-sentence c_2, in effect, correlates the points in the positive segment with the positive real numbers, since each finite positive segment l will determine a unique point p in the x-axis, namely the end-point of l that is not o.

With the one-one correspondence established between the points in the positive part of the x-axis and the positive real numbers, it is a simple matter to set up the analogous correspondence between the points in the negative part of the x-axis and the negative real numbers. Continuing in this vein, we can specify a one-one correspondence between the points in a plane and the ordered pairs of real numbers.

Since functions can be viewed as sets of ordered pairs, there is clearly no difficulty in developing, within the framework of the constructibility theory, the theory of functions of a real variable. All the standard theorems can be obtained and the usual applications of the theory can be made. In particular, we now have a theoretical basis for representing geometrically, the graph of a function of a real variable, and we can explain why the definite integral $\int_a^b f(x)dx$ of a continuous function that is everywhere positive in the interval $[a, b]$ gives one the area bounded by a, b, the x-axis, and the curve determined by $f(x)$.

4. Putnam's Case for Realism

Newton's Law of Universal Gravitation seems to make essential reference to such mathematical entities as numbers and functions: it says roughly that "bodies behave in such a way that the quotient of two numbers *associated* with the bodies is equal to a third number *associated* with the bodies" (Putnam (What), p. 74). (The numbers being talked about here are real numbers.) Because of this, Hilary Putnam once claimed that denying the existence of numbers and functions, while at the same time accepting as true (or even approximately true) this law of Newtonian physics, "is like trying to maintain that God does not exist and angels do not exist while maintaining at the very same time that it is an objective fact that God has put an angel in charge of each star and the angels in charge of each of a pair of binary stars were always created at the same time!" (p. 74).

I shall say more about Putnam's reasoning in Chapter 8; but for now, a few initial remarks are in order. One can analyse away such apparent references to mathematical objects in the way outlined above. We have seen how reasoning involving numbers and functions can be carried out within the constructibility theory, without quantifying over numbers and functions. Indeed, since it is clear that the mathematics of Newtonian gravitational theory can be developed within Simple Type Theory, the constructibility theory provides us with the means of expressing the empirical content of this physical theory. Thus, the acceptance of the truth (or approximate truth) of Newton's Law of Universal Gravitation does not automatically carry a commitment to the existence of mathematical objects.[1]

According to Putnam, the whole study of space in modern mathematics is based on the postulate that there is a one-one correspondence between the real numbers and the points on a line (What, p. 64). When this postulate was found to be extremely fruitful both in mathematics and in physics, scientists were justified in accepting the postulate. The postulate had been given what Putnam calls a "quasi-empirical" justification. It can be seen, however, that the quasi-empirical justification of the correspondence postulate should not lead one to believe in the actual existence of such mathematical entities as the real numbers. For the theoretical benefits of accepting the correspondence postulate can be obtained, in good measure, by accepting instead the one-one correspondence, described in the previous section, between what was described as "the points on a line" and "the real numbers". Of course, under analysis, the "points"

---

[1] Putnam also suggested that the acceptance of the Law of Universal Gravitation requires belief in the existence of such measurable quantities as distances, masses, and forces. It would take me too far afield to explore such a belief here. Suffice it to say that I have shown how the apparent reference to some measurable quantities can be analysed away. For example, one does say, in physics, that the torque generated by applying a force to a lever arm is equal to the force exerted on the lever arm times the length of the lever arm. Does that mean that, in addition to the things that have lengths, there are entities that are lengths? Not at all. As I showed earlier in this chapter, talk about lengths can be taken to be a mere manner of speaking about things that have lengths. For many of the measurable quantities discussed in physics, similar analyses can be given, but this is not an appropriate place to carry out such analyses.

and the "numbers" talked about in the constructibility version of the correspondence postulate are mere fictions.

5. Complex Analysis

I turn now to a brief consideration of that especially beautiful and useful branch of classical analysis known as the *"theory of functions of a complex variable"*—a branch of mathematics that is heavily used in science and engineering.

Complex numbers were first introduced in the theory of cubic equations (Struik (Concise), p. 114). Initially, they were thought to be somewhat mysterious. In particular, i, the square root of -1, was puzzling to some: after all, how could there be a number whose square is -1? (Thus, the term "imaginary number".) However, much of the mystery surrounding the idea of complex numbers evaporates when it is seen that complex numbers can be regarded simply as ordered pairs of real numbers. Thus, the complex number $z = x + iy$ can be taken to be the ordered pair $<x, y>$. Addition of complex numbers can then be defined so that the sum of complex numbers u and v is that ordered pair whose first component is the sum of the first components of u and v and whose second component is the sum of the second components of u and v. Similarly, the operations on complex numbers of multiplication, subtraction, and division can be defined in the obvious ways in terms of these ordered pairs. Thus, starting with the real numbers—here, we can take as real numbers the constructibility versions of the Dedekind cuts defined in the standard way from the finite cardinals—it would be a simple matter to construct, within the constructibility theory, essentially the standard theory of complex numbers and the theory of functions of a complex variable: all the standard theorems can be obtained and the usual applications of the theory, such as the use of conformal mappings to solve boundary value problems in partial differential equations, can be made.[2]

Some idea of how this theory is applied in science and engineering can be obtained from the following considerations: With only minor changes in what was described in the previous section, complex numbers can be represented geometrically as points in a plane. Now suppose that we represent complex numbers, $x + iy$, by plotting

[2] The reader can consult R. Churchill (Complex) for more details.

values of x and y as rectangular coordinates in a plane. Let us call this xy plane "the *complex plane*". Let $w = f(z)$ be a function of a complex variable z. Then, corresponding to each point $<x, y>$ in the complex plane from which $f(x + iy)$ is defined, there will correspond a complex number $u + iv = f(x + iy)$. Clearly, these values can themselves be regarded as points in a plane. Let us represent these corresponding values as points in a new plane, the uv plane. Then, the function $f(z)$ maps points in the complex plane to points in the uv plane. A study of such mappings (or transformations) has proved to be remarkably fruitful to scientists and engineers. For example, for purposes of solving certain kinds of boundary value problems in Laplace's differential equations, it has been found useful to investigate the way in which various regions are transformed by conformal mappings. (See Churchill (Complex), ch. 9, for details.) These applications can be readily understood within the framework of the constructibility theory.

6. Applications by Means of Structural Identity

Let us return now to a topic discussed in the opening sections of this chapter: the real number system. We saw earlier that essentially the standard Dedekind cut method of constructing real numbers can be carried out in the constructibility theory. This system of positive real numbers (with $+$ and $\times$) can be easily seen to be isomorphic to the system of real lengths described above. So instead of carrying out a detailed development of a theory of functions of real lengths, we can make use of the standard theory of functions of the positive real numbers. This is because the isomorphism will allow us to pass back and forth between the two systems, allowing us to draw all the conclusions about functions of real lengths that we need to draw from our conclusions about functions of positive real numbers, without having to utilize a special theory of functions of real lengths.

This example brings out a significant feature of standard mathematical practice. Suppose that what was most important mathematically to physicists were theories of functions of measurable quantities. Even so, mathematicians, for the most part, would not bother to work with measurable quantities: the structure of the real numbers underlying the system of measurable quantities would be abstracted and the mathematician would tend to work with the abstracted system. As a general rule, mathematicians concern themselves only with those features of a system that are "mathematically significant" and these tend to be structural features. No doubt, facts such as the above lend plausibility to the philosophical position known as 'Mathematical Structuralism' (which I shall discuss in detail in the following chapter), according to which mathematics is the study of structures. But it needs to be kept in mind, in order to counteract the seductive simplicity of the Structuralist's model of applications of mathematics in physics, that there is no necessity for applying mathematics in the way sketched above, whereby one first abstracts the essential mathematical structure of the real number system from the systems of real magnitudes, and then theorizes about the mathematical structure, returning when appropriate to the system of magnitudes by way of isomorphism. One could remain within the systems of magnitudes and do one's mathematics in such a system.

Part II
PHILOSOPHICAL DEVELOPMENTS

IN the second part of this work, rival philosophical views are considered and evaluated in the light of the constructibility theory. The philosophical outlook of the author emerges by way of contrast with the rival views. Replies to anticipated objections are presented, and various details of the author's philosophical position are clarified.

7

Mathematical Structuralism

1. Introduction

THE famous mathematician Nicholas Bourbaki, writes in the conclusion of his article 'Architecture of Mathematics':

[M]athematics appears thus as a storehouse of abstract forms—the mathematical structures; and it so happens—without our knowing why—that certain aspects of empirical reality fit themselves into these forms, as if through a kind of preadaptation. . . . The unity which it [i.e. structure] gives to mathematics is not the armor of formal logic, the unity of a lifeless skeleton; it is the nutritive fluid of an organism at the height of its development, the supply and fertile research instrument to which all the great mathematical thinkers since Gauss have contributed (Architecture, p. 231).

We have here an expression of a view of the nature of mathematics that has been classified as a Structuralist view. Others who are thought to have "articulated and defended Structuralist philosophies of mathematics" are M. Barbut, P. Benacerraf, R. Dedekind, M. Steiner, and M. Wilson.[1] In this chapter, I develop and also clarify the overall view of mathematics of the present work by contrasting it with the views of some Mathematical Structuralists. In particular, I discuss the works of two philosophers who have been most explicit in their espousal of Mathematical Structuralism.[2]

2. The Structuralist View of Applications

In (M and R), Stewart Shapiro describes Mathematical Structuralism as the doctrine that the subject-matter of mathematics consists of

---

[1] See Stewart Shapiro (M and R, pp. 535–6).

[2] The two philosophers whose views on Structuralism I discuss in detail in this chapter, Stewart Shapiro and Michael Resnik, have worked closely together. However, the reader should be cautioned against assuming that they form a team or that they agree with one another on all major points with respect to Structuralism.

structures (or patterns). What are structures? As a first approximation, one can think of a structure as a domain of objects together with
some relations or functions (or both) that satisfy some given
conditions. In other words, one can take Shapiro's structures to be
the familiar structures of mathematical logic.

But the familiar structures of logic are usually defined set-
theoretically: they are generally taken to be set-theoretical objects of
some sort. The Structuralist wants to place set theory on a par with
the other branches of mathematics. Set theory, for Shapiro, does not
provide us with the foundational structure of all the other branches of
mathematics; set theory merely studies one of many possible
structures. Arithmetic, he tells us, is not the study of a particular set of
things called 'natural numbers'; rather, it is the study of "the natural
number structure"—"the structure or pattern of any system that has
an infinite number of objects with an initial object and a successor
relation (or operation)" (M and R, p. 534). Hence, we need a
characterization of the notion of structure that is not set-theoretical.

Shapiro doesn't attempt any sort of definition or exact characterization. Instead, he gives some examples and gives hints for grasping
structures. He does put forward the suggestion that a mathematical
structure might be construed as "the form of a possible system of
related objects, *ignoring the features of the objects that are not relevant
to the interrelations*" (M and R, p. 535, italics mine). Thus, consider
the structure of Zermelo–Fraenkel set theory. This gives the
membership relationships among the objects in the structure (i.e.
among the sets), but all features of these objects that are not relevant
to the interrelations are to be ignored, i.e. we ignore everything other
than the membership relationship, and we ignore the membership
relationships that may obtain between the objects and anything
outside the structure (e.g. if there are sets of urelements—sets of
physical objects, say—these relationships are to be ignored).

Shapiro's idea that the subject-matter of mathematics consists of
structures can be illustrated by turning to algebra. It can be plausibly
argued that algebra is the study of algebraic structures: groups, semi-
groups, rings, integral domains, fields, etc. Paul Benacerraf once
proposed a doctrine similar to the above, claiming that for
arithmetical purposes, "the properties of numbers which do not stem
from the relations they bear to one another in virtue of being arranged
in a progression are of no consequence whatsoever" (Numbers,
p. 291). And he went on to argue:

That a system of objects exhibits the structure of the integers implies that the elements of that system have some properties not dependent on structure. It must be possible to individuate those objects independently of the role they play in that structure. But this is precisely what cannot be done with the numbers. To *be* the number 3 is no more and no less than to be preceded by 2, 1, and possibly 0, and to be followed by 4, 5, and so forth. . . .

Arithmetic is therefore the science that elaborates the structure that all progressions have in common merely in virtue of being progressions (p. 291).[3]

Shapiro suggests that there are two main advantages to Structuralism (M and R, pp. 541–2):

(1) It provides a more holistic view of mathematics and science, and thus accounts for the rich interplay between the fields. The main idea here is that we can view scientific theories as incorporating well-developed mathematical structures. Because of this, the mathematician who proves theorems about the structures gives the scientist information about that aspect of reality being theorized about. Furthermore, empirical discoveries may provide the mathematician with information about some mathematical structure.

(2) It accounts for the interconnections among the various branches of mathematics. Mathematical structures frequently share common substructures. For example, the real number structure "contains" the natural number structure as a part. Thus, theorizing about one structure may provide us with information about another; and we can see how real analysis could be used to prove theorems about the natural numbers.

What is the Structuralist's explanation of the relationship between mathematical knowledge and scientific knowledge, and also of applications of mathematics in science? Crudely put, the explanation runs (M and R, pp. 538–9): (1) The scientist finds some mathematical structure exemplified in the physical world. (2) The mathematician provides detailed information to the scientist about this structure. (3) This information is then used by the scientist to infer facts about the physical system of objects that is an exemplification of the structure. Shapiro admits that the above account is overly simple and hastens to add a correction. Consider the case of the use of Euclidean geometry in physics. Did the physicist actually discover a system of physical

---

[3] So far as I know, Benacerraf has not developed these ideas into anything like a Structuralist philosophy of mathematics. For criticisms of Benacerraf's reasoning in this paper, see M. Steiner (Mathematical) and my (Ramsey).

objects that exemplifies this geometric structure? The problem is with the space points. Did the physicist discover the continuum of points in space? With what instruments could he or she discover such points? Shapiro suggests that the discovery of the exemplification of the Euclidean structure was indirect and involves the postulation of theoretical entities, i.e. we are to think of the space points as analogous to such theoretical entities as molecules and electrons— postulated to yield adequate theory and explanation (M and R, pp. 538–9). But with this correction, the above account of how mathematics is applied in science is essentially the Structuralist's account.

The following quotation from a work by the physicist Robert Geroch seems to support Shapiro's view of applications:

> What one often tries to do in mathematics is to isolate some given structure for concentrated, individual study: what constructions, what results, what definitions, what relationships are available in the presence of a certain mathematical structure—and only that structure? But this is exactly the sort of thing that can be useful in physics, for, in a given physical application, some particular mathematical structure becomes available naturally, namely, that which arises from the physics of the problem. . . . The idea is to isolate mathematical structures, one at a time, to learn what they are and what they can do. Such a body of knowledge, once established, can then be called upon whenever it makes contact with the physics (Physics, p. 1).

I shall return to an examination of this quotation in a later section.

Let us consider the following objection to the Structuralist's view. According to Shapiro, mathematical systems are sometimes applied in physics even though there are terms in the system that do not stand for any physical objects or relations among physical objects. For example, he cites the use of the theory of functions of complex variables in physics, and allows that the complex number terms do not refer to any physical objects. Now this admission may be thought to create a difficulty for Shapiro's account of how mathematics is applied in physics, for the account seems to require that the physicist discover in nature an exemplification of the mathematical structure to be applied. This suggests that he must find *physical* objects and relations among these objects which together form a configuration that is an exemplification of the structure in question. But what physical objects will stand in the places in the exemplification that correspond to those occupied in the mathematical structure by the complex numbers?[4] So far as I know, Shapiro has not worked out a satisfactory response to this objection. One possible way of responding occurs to me, which may in the end turn out to have some plausibility. Shapiro might appeal to the fact that it is sometimes useful, in attempting to obtain information about one structure, to work in a richer structure in which the structure in question can be imbedded. Thus, the fundamental theorem of algebra requires a difficult and complex proof when one sticks to purely algebraic methods, but it can be given a very simple and elegant proof by working in the theory of functions of complex variables (see, for example, Churchill's (Complex), p. 96). To take a simpler example, suppose that one needs a quick rough estimate of the sum

$$\sum_{n=1}^{100} n^{17}$$

This can be obtained rather easily if one makes use of real analysis. In particular, it is easy to see that the definite integral

$$\int_0^{100} x^{17}\, dx$$

provides an easily calculable estimate. Thus, Shapiro could respond to this objection as follows:

I did not claim that every mathematical structure that gets applied must have a physical exemplification. If the complex number system gets applied in physics, then according to my account, there must be *some* mathematical structure that is found to be exemplified in the physical world by the physicist, but it need not be the structure of the complex number system itself. Suppose, for example, that the structure $R \times R \times R$ (where $R =$ the real numbers) is exemplified in the physical world by the continuum of points in space. Now the complex number structure is a substructure of various mathematical structures in which $R \times R \times R$ is also to be found as a substructure. Indeed, these two substructures are interrelated in various ways: R, for example, is a substructure of the complex numbers. I suggest that *the complex number structure is*

---

[4] This objection was suggested to me by Paul Teller and Elijah Millgram in a graduate seminar I gave in the fall of 1984.

used by scientists to infer facts about $R \times R \times R$ (or some encompassing structure that is also physically exemplified).[5]

However, before this response would be at all convincing, an investigation would have to be carried out to see if all scientifically useful applications of the complex number structure in science are useful for the scientist *because of the information that is provided about* $R \times R \times R$ (or some other physically exemplified structure).

Let us attempt to become clearer about the position I have sketched above. According to this Structuralist account of how mathematics is applied to give us knowledge about the physical world, an application of a mathematical theory presupposes the discovery that the mathematical structure of this theory is exemplified in "material reality". The idea is: in using mathematics to draw some conclusion about some particular phenomenon, one must know that the phenomenon is embedded in a structure that is an exemplification of a mathematical structure. But is this required of *each* person who applies this mathematical theory? And is this required of *each* application of the mathematical theory? Evidently the answer to both these questions is "Yes".

To see why I interpret Shapiro in this way, consider how he responds to the following objection. The structure of number theory is infinite. This suggests that "*whenever* one uses numbers or arithmetic in everyday life, such as counting tomatoes or balancing a checkbook, one is presupposing an infinite structure" (M and R, p. 541, italics mine). But it seems absurd to suppose that a child, who has no understanding of infinity,[6] can be applying arithmetic in this way. Now what is striking about Shapiro's reply to this objection is not so much what he says, but what he does not say. Shapiro does not claim that this objection in any way misinterprets his view: he does not say that it is not *each* application of mathematics that presupposes a knowledge that the structure in question is exemplified in material reality. Instead, he replies that "for virtually all applications of arithmetic, one does not require the entire natural

---

[5] The passage concerning the above objection, as well as this reply made on behalf of Shapiro, were sent to Shapiro for comment. Nothing he wrote in his letter of response suggests that he has any disagreement with this account of his views.

[6] My daughter, at the age of 7, once came to me with her homework assignment: "I can't do this one," she said, pointing to a division problem. It was: eight divided by zero! (I assume that the teacher simply made a slip.) "What number gives eight when it is multiplied by zero?" she wondered out loud. "It must be very big . . . VERY BIG." After a pause, she asked: "Could it be ninety?"

number structure, but only various finite structures which can be seen as 'part of' the natural number structure" (p. 541). Shapiro regards the child who learns to count to twenty as acquiring knowledge of a finite structure involving a twenty-element sequence. Thus, in order to apply the arithmetical knowledge that has been acquired, it is sufficient that the child see that the finite structure is exemplified in the physical world: it is not required that the child see that the full infinite structure of the natural number system is exemplified.

It would seem then that, according to Shapiro's view, each and every application made of mathematics is to be understood in the way described above. But he does not maintain that every application of a mathematical theory involves seeing that the whole structure described by the theory is exemplified in material reality. Application requires only that part of this structure be seen to be exemplified.

3. Resnik on the Nature of Structures

Let us now examine more closely the notion of structure or pattern that underlies this philosophy of mathematics. For a discussion of this fundamental idea and of the ontological status of structures, I turn to the writings of Michael Resnik, since his descriptions of these items have been the most detailed.[7] In (MaSoP:O), Resnik describes two philosophical difficulties that arise out of the traditional Platonic view of mathematics, according to which mathematics describes abstract objects that do not exist in physical space or time. The first difficulty is to explain how we can have knowledge of such objects. For if we, the knowers, exist in physical space and time, whereas the objects known do not, how can we interact causally with these objects? And if we cannot causally interact with these things, how can we possibly obtain any knowledge of them? The second difficulty is

[7] Shapiro is much less definite about the nature of structures. In fact, unlike Resnik, Shapiro does not commit himself to a view about the ontological nature of structures. He espouses not ontological Platonism but rather "methodological Platonism". He claims he is a methodological Platonist in so far as he uses the word 'structure' as a common noun and quantifies over structures. I should also note that Resnik's own view of applied mathematics differs somewhat from the view I have attributed to Shapiro. Resnik believes that a physical theory is, for the most part, an applied *mathematical* theory, i.e. a theory of a structure, some of whose elements are abstract (mathematical) entities. He thus does not subscribe, in an unqualified way, to Shapiro's picture of the physicist finding concrete physical realizations of mathematical structures.

due to the fact that no mathematical theory can do more than determine its objects up to isomorphism. This places the Platonist in the awkward position of claiming both that a given mathematical theory is about certain things and also that it is impossible to state definitely what these things are. Resnik claims, however, that these problems are due to a "fundamental misconception of what mathematics is about":

If we conceive of the numbers, say, as objects each one of which can be given to us in isolation from the others as we think of say, chairs or automobiles, then it is difficult to avoid conceiving of knowledge of a number as dependent upon some sort of interaction between us and that number. The same line of thought leads us to think that the identity of a number *vis-à-vis* any other object should be completely determined (p. 529).

But, according to Resnik, we are never "given" any mathematical object in isolation but only in structures: "That 13 is a prime number is not determined by some internal property of 13 but rather by its place in the structure of the natural numbers." For Resnik, the objects of mathematics, i.e. the entities which the terms and quantifiers of our mathematical system denote, are structureless points or positions in structures. Furthermore, these structureless points have "no identity or features outside of a structure" (p. 530).[8]

But what is a structure? For Resnik, a structure is "a complex entity consisting of one or more objects [which he calls 'positions'] . . . standing in various relationships" (p. 532). Two structures are said to be *congruent* or *structurally isomorphic* if there is a one-to-one correspondence between the positions of the structures preserving all the operations, relations, and distinguished positions of the structures (such as zero in the natural number structure). An arrangement of concrete objects instantiates a structure if it is congruent to that structure. Resnik goes on to define various other relationships that structures may enter into: Structures may *occur in* other structures; and structures can be *definitionally equivalent*. But these details need not detain us here.

There are two basic sorts of objects that Resnik talks about: (*a*)

[8] The reader may wonder how this structuralist view is supposed to solve the first of the two problems described above. I have doubts that this radical view gets the Platonist very far towards anything like a real solution, and in any case, I can say that Resnik does not make his "solution" to the problem at all clear. However, the reader should study Resnik (MaSoP:E) for his most serious attempt to date to deal with the epistemological problems of Platonism.

structures or "patterns", as he calls them, and (*b*) objects (or positions) in structures. Both kinds of "objects" are very strange. For according to Resnik, it makes no sense to say of structure *A* and structure *B* that *A* = *B* or that *A* is not identical to *B*, i.e. structures are excluded from the field of the identity relation. And although he allows that identity may or may not hold of objects within a single structure, when it comes to objects in different structures, identity has no role to play (MaSoP:O, pp. 536–8). If *x* and *y* are objects from different structures, you not only cannot say that *x* is identical to *y*, you cannot say that *x* is different from *y* either. This is what I shall call 'the doctrine of the nonsensicality of trans-structural identity'. But wait. How can I talk of different structures in the above? For doesn't 'different' mean 'not identical to'? And how can I say *that*? For, as I noted above, according to Resnik we cannot say of structure *A* that it is or is not identical to structure *B*.

Now how did Resnik get into this paradoxical position? I have learned from Resnik that the view was a result of "reflecting on what it would take for there to be a fact of the matter as to the identity of positions in patterns that we do not take to be parts of a larger pattern".[9] He was moved to conclude that there simply is no such fact of the matter. He was then led to consider what is sometimes called "Benacerraf's multiple reduction problem":[10] within the structure of some standard set theory, one can discern many substructures that are congruent to the natural number structure (omega, +, ×). So this multiplicity of structure gives rise to the question: Which set in the set-theoretical structure is the natural number 2? It would seem that any answer one gives will be arbitrary. Resnik, however, thinks he has a kind of answer to this problem. It makes no sense, he claims, to ask such a question: across structures, one cannot ask if an object in one structure is or is not identical to an object in another structure. Thus, the doctrine of the nonsensicality of trans-structural identity is brought in to take care of Benacerraf's problem (MaSoP:O, pp. 541–2). But by adopting this doctrine, Resnik puts himself in an awkward position regarding structures; for suppose that it is asked what it means to say of structure *A* that it is or is not identical to structure *B*. Well, we would like to say that structure *A* = structure *B* if, and only if, every position in *A* is a position in *B* and . . . Whoops!

[9] This I learned from a personal communication from Resnik.
[10] See Benacerraf (Numbers). Cf. also Kitcher (Plight).

We are assuming that it makes sense to say of a position in *A* that it is (identical to) a position in *B*. So it is easy to see how Resnik came to exclude structures as well as positions from the field of identity.

Let us backtrack a bit. Having assumed that natural numbers are objects in a structure, Resnik tries to avoid Benacerraf's problem by simply denying that it makes sense to say of the natural number 2 that it is or is not identical to such-and-such set. But the question arises: Why does Resnik assume that natural numbers are objects at all? That is, why does Resnik take the position that there is some object that is the number 2? To answer this, we need to re-examine the very first paragraph of his paper, where it is said: "I seek an account of mathematics in which the logical form of mathematical statements are taken at face value and their semantics is standardly referential, say, in the manner of Tarski." Now the "standard referential semantics" of, say, first-order logic proceeds by specifying the truth-conditions of the sentences in terms of reference: these names stand for these things, and these predicates are to be true of (or satisfied by) those kinds of things, etc. And the things that a predicate is true of are those *existing* things that are such-and-such. (A predicate cannot be true of something that does not exist.) Similarly, the terms and names cannot stand for things that do not exist. (Of course, an expression might seem to name something, when in fact it does not.) Now Resnik is happy with this sort of account. But if we take the logical form of mathematical statements at face value, the standard referential theory would seem to require that there be (i.e. exist) mathematical objects for the mathematical terms to denote and for the mathematical predicates to be true of. Thus, it would seem, he is led to his doctrine that the objects in a structure truly exist.

But why suppose that the language of mathematics ought to be analysed in the above referential way? Footnote 1 of Resnik's paper indicates the sort of consideration that prompted Resnik to adopt the referential view. There, he cites with approval Benacerraf's argument in (Truth) for giving a uniform semantics for mathematics and the rest of science: "[I]f mathematical terms have one kind of semantics while other scientific terms have another then a special semantics will still be required for sentences containing both kinds of terms." The suggestion is that since the language of science has, supposedly, been shown to have the standard referential semantics, we have strong motivation for analysing the language of mathematics to have the same referential semantics.[11] Resnik goes on to argue that this argument can be strengthened by noting that "a uniform semantics for both mathematics and science will be required to obtain an account of inference involving both mathematical and scientific sentences."

4. An Evaluation of Shapiro's Account of Mathematics

I turn now to a critical examination of the Structuralist's doctrines, separating my discussion into two parts. In the first, I shall take up Shapiro's views about the nature of mathematical knowledge and about how mathematical knowledge is applied in the empirical sciences. Then in the second part, I shall discuss Resnik's ontological views about structures.

Shapiro's theory of how mathematical knowledge gets applied in science basically comes to this: (1) The mathematician gives us knowledge of mathematical structures; (2) the scientist discovers among observable (or theoretical) physical objects an exemplification of a structure; and (3) the mathematician's knowledge of the relevant structure is then used to draw conclusions about this exemplification of the structure. What is striking about Shapiro's position is that this is the *only* way in which mathematics gets applied in the physical world. Now there is a danger that a monolithic view of this sort may be made so flexible in order to escape counter-examples that it ends up having practically no content at all. I am reminded of Wittgenstein's example of the astronomer who first claims that all astronomical bodies move in circles, and then later, to deal with data contradicting his claim, says that all astronomical bodies move in circles *with deviations*. So long as the notion of "deviation" is left unspecified, it is not clear what content we should attribute to such a claim; for any trajectory in space can be described as a deviation from a circular path. If Shapiro's views are not to be empty, something definite and determinable must be meant by 'discover among observable (or theoretical) physical objects an exemplification of a structure'. Shapiro attempts to provide some content to this phrase by saying: "The claim that a structure is exemplified in physical reality

---

[11] Cf. my (V-C) ch. 3, sect. 4, on Quine's position regarding the language of science.

amounts to a claim that there are entities of some sort that answer to the places within the structure" (M and R, p. 539). Thus, recall the example of the complex number system: it was thought that this system was not so exemplified because there seem to be no physical objects corresponding to the "imaginary numbers". On the other hand, in order to have a plausible physical exemplification of Euclidean geometry, Shapiro requires us to postulate a continuum of space points, claiming that "space points are theoretical entities called for by the use of Euclidian geometry in physical theory" (p. 540).

But how plausible is this view that people in fact postulate space points? On what grounds does Shapiro attribute this postulation to ordinary people applying Euclidean geometry in the most mundane situations? Does the ordinary man in the street believe that, among the physical objects we ought to believe in, there are these unobservable ones that the geometer calls 'space points'? Surely not. It would seem that Shapiro has people making such postulations not because he has conducted an investigation into how people in fact reason, but because his monolithic view of how mathematics gets applied requires such postulations.

In support of the above points, I would like to revert to Euclid's version of geometry. As I noted in Chapter 3, Section 4, this was a theory that talked about what geometric constructions are possible —not about what geometric entities exist. What one finds among the axioms of Euclid's geometry are not existential assertions but rather assertions about what it is possible to construct. Thus, the third postulate tells us that it is possible to construct a circle with a given centre so as to pass through a given point. Although Euclid's geometry can be transformed relatively easily into a theory of geometric structure of the sort Shapiro has in mind, it would be misleading to describe Euclid's theory as such a theory. Now the question I should like to consider here is this: In applying Euclid's geometry, must one postulate the actual existence of space points in order to have a physical exemplification of the mathematical structure that is described by modern versions of the theory? The answer is clear. It is possible to apply this geometry without even knowing what mathematical structure underlies this geometry. A schoolboy, for instance, who proves the standard theorems in such a system, could apply these theorems to draw conclusions about the lines marking off fields of wheat, without having to postulate space points or to see any actual physical exemplifications of Euclidean structures. There are several reasons for maintaining that such a possibility exists. First of all, the lines marking off fields of wheat can be regarded as at least approximations of lines that Euclid's geometry tells us can be constructed. So the conclusions that can be drawn in the geometry about such lines give us information about the boundary lines straightforwardly, without requiring a circuitous chain of reasoning by structural isomorphism. In short, one can see how this geometry can be applied without the need for the special kind of postulation Shapiro's theory of application requires. Secondly, recall that Euclid's geometry deals with what points and lines it is possible to construct. Now there is a sort of absurdity in supposing that the people who constructed this theory believed in the actual existence of the things that the theory says *could* exist. If they believed that the things already existed, why did they frame their theory to say merely that these things can be constructed? Besides, why would they feel the need to postulate such things? If I wish to apply a theory of what buildings can be built in a certain location in San Francisco, must I postulate the actual existence of the buildings that can be built? Thus, it is implausible to suppose that when this geometry was used in Ancient Greece to solve practical and scientific problems, it was felt necessary to postulate a continuum of actually existing space points. We seem to have here a mathematical theory that not only can be applied without first finding an exemplification among physical objects in the way Shapiro's theory requires, but it in fact was so applied.

In connection with this last point, I should like the reader to reconsider the constructibility theory developed in this work. Can we not imagine someone using such a theory to reason about finite cardinality in the way sketched in Chapter 5? Indeed, might not this person draw all the right number-theoretic inferences needed in everyday life without ever talking about natural numbers, sets, or mathematical objects of any sort? Indeed, need this person have in mind any mathematical structure at all? Could he or she not just use the logic of this system to draw the right conclusions? For all we know, some ancient Babylonian mathematicians may have reasoned in ways closer to what is being envisaged here than to what Shapiro describes.

Of course, it could be replied that this imagined theorizer is not truly applying *mathematics*—it just looks as if he or she is. But such a reply would not be plausible. For the "set theory" of the constructibility theory is indistinguishable from the set theory of the usual Simple Theory of Types. It is hard to see how one could reasonably maintain that applications of the latter count as genuine applications of mathematics, if one is to deny that applications of the former count as applications of mathematics. What would be the rationale for such a strange position?

The Structuralist sees the empirical scientist going to nature with a bag full of mathematical structures, exemplifications of which he or she tries to find among the physical objects. Although I would not deny that something like this happens some of the time, I see no reason to suppose that this sort of thing happens whenever mathematics is applied.

For purposes of contrast, I wish to propose an alternative model. The Structuralist's model of application is avowedly Platonic in conception: structures are like Plato's Forms, and the physical exemplifications are like Plato's sensible objects that resemble or "partake of" the Forms. Ponder the following Kantian model, according to which applications of mathematics unfold out of the nature of the fundamental concepts with which the scientist attempts to understand nature—especially those fundamental concepts underlying the scientist's ideas about the comparison and measurement of physical things. Thus, imagine a scientist beginning an analysis with some rather vague and indefinite concepts of 'having length', 'being equally long as', 'being straight', 'being a part of', and so on, which are used in developing the idea of comparing all objects (of the appropriate sort) with respect to length. In the process, some of the ideas used in the analysis are sharpened, simplified, or idealized, but eventually, by straightforward reasoning, and without ascending to an overview of a structure, the scientist arrives at the concept of an object being n/m as long as another, and is able to discover and, in some cases, even deduce such laws of combination and comparison as are reflected in the usual laws of addition and multiplication of rationals. Now as the theory of combinations and comparisons of objects with respect to length gets developed, a structure may slowly emerge. So we can imagine applications of the theory along the lines sketched by Shapiro, i.e. we can imagine someone finding an exemplification of such a structure and drawing conclusions about the exemplification on the basis of what is known about the structure. But it is clear that applications of the theory need not proceed in such a way. The mathematical theory developed could be applied directly to practical questions involving the lengths of physical objects and boundary lines, without the need for the kind of reasoning by way of structural isomorphism that is essential to the Structuralist's model of applications. To see how this can be done, the reader should review the treatment of rational and real lengths given in the previous chapter.

I do not wish to suggest that, in fact, the use of rational number theory in calculating and comparing lengths, areas, volumes, and the like came about in this way. I am sure that the actual development of our techniques of applying rational number theory to the world was much more complex and involved than is suggested by either of the models being discussed here. I suspect that this development involved, among other things, much trial and error, shrewd guesses, reasoning by analogy, and extrapolations from previously successful procedures and techniques. However, I believe it is worth while having some simple idealized models of how such techniques could have developed if only to appreciate more fully the complexities of the actual historical development.

It is curious that Shapiro does not discuss in (M and R) the semantical approach to scientific theories advocated by Patrick Suppes, Bas van Fraassen, Fredrick Suppe, and many others.[12] For this approach emphasizes structures instead of sentences. Under the semantic view, theories are characterized by specifying a class of mathematical structures to be used for the representation of empirical phenomena; so one would have thought that Shapiro would have been sympathetic with such an approach. The semantic approach, however, suggests that applications of mathematical structures to the physical world are frequently not made in the straightforward way described by Shapiro. For example, in evolutionary theory, structures are frequently used to describe *ideal* populations of organisms.[13] In such cases, *there may be no actual population that exemplifies the structure.* But the idealization may be useful just the same in making certain kinds of estimates, in designing experiments, in making predictions, in explaining certain features of actual population growth, etc.

I would now like to return to the quotation from Geroch I gave in Section 2. Recall that Geroch described the use of mathematics by

---

[12] See Suppes (Scientific), van Fraassen (Semantics) and (Image), and Suppe (Structure) for expositions and defences of this approach.

[13] See E. Lloyd (Evolutionary, ch. 2, sect. 2.3 for details).

physicists in a way that strongly suggested the Shapiro view. The idea was that the knowledge obtained by mathematicians about some structure could be "called upon whenever it makes contact with the physics". The question is: *How does the mathematical structure make contact with the physical system in typical applications of mathematics*? Let us ponder the following example from (Physics) of an application of topology in the analysis of some classical dynamic system:

We imagine some physical system that we wish to study. It is perhaps difficult to say exactly what one means by "a physical system," but we imagine it as consisting of some mechanism which sits in a box on a table, unaffected by "uncontrollable external influences." We shall, however, allow ourselves to influence the system (e.g., by hitting it with a stick) so that we may manipulate the system to study its properties. We introduce the notion of "the state of the system," where we think of the state of the system as a complete description of what every part of the system is like at a given instant of time. (For example, the state of a harmonic oscillator is specified by giving its position and momentum.) Thus, at each instant of time, the system is in a certain state, and all we can ever hope to know about the system, at a given instant, is what state it is in. By an extended series of manipulations on the system, we "discover all the states which are available to it," that is, we introduce a set Γ whose points represent the states of our system. Thus our mathematical model of the system so far consists simply of a certain set Γ.

We next decide that, more or less, we know what it means physically to say that two states of the system (i.e., two points of Γ) are "nearby". Roughly speaking, two states are "nearby" if "the system does not have to change all that much in passing from the first state to the second." We wish to incorporate this physical idea as mathematical structure on the set Γ. The notion of a topology seems to serve this purpose well. Thus we suppose that the set of states, Γ, of our system is in fact a topological space, where the topology on Γ reflects "physical closeness" of states (Physics, pp. 183–4).

Now does this example fit the Shapiro model? What are the physical entities that make up the physical exemplification of the mathematical structure? The points in the topological space correspond to "states", and "states" are regarded as descriptions, not physical objects. And if there are infinitely many of these "states", then the descriptions do not even exist in the physical world, since presumably infinitely many (tokens of) descriptions have never been constructed. I conclude that the application of mathematics described above does not conform to Shapiro's description of how mathematics is applied in science.

It should be pointed out that this dynamical system can be

described within the framework of the constructibility theory. Starting with the non-mathematical description of the physical system, statements about the existence of "states" can be construed as statements about the constructibility of the descriptive open-sentences. As for the mathematical system needed, the notions of topological space and topology are straightforwardly set-theoretical notions, so they can be defined within the constructibility theory. How about other features of the mathematics of this system? Geroch suggests that we "think of an observable as an instrument with a dial such that, when the instrument is brought into contact with the system, the dial reads a certain real number" (Physics, p. 187). This idea gets translated into the mathematics as a continuous, real-valued, function on Γ. I showed in Chapter 6 how the real numbers can be treated in the constructibility theory. Thus, the function needed by the above idea can be defined within the constructibility theory. Next, consider the way in which the dynamics of the system is described. "It is observed, in the physical world, that prediction is normally possible for physical systems: if you tell me the state of the system now and how long you are going to wait (and if you do not interfere with the system during its evolution), then I can tell you what state the system will be in at the end of that time" (Physics, p. 185). This picture gets incorporated into the mathematical description by postulating a continuous function $f(x,t)$ which takes point x and time t to a point y. In short, $f(x,t)$ represents the state in which the system will be after time t, if it was initially in state x. Thus, the dynamics of the system get described by a continuous mapping of $\Gamma \times R$ to Γ. Here again, there is no problem in doing all of this mathematics within the constructibility theory: the mapping needed can obviously be "found" within the system. Finally, Geroch suggests that the topology on Γ may be used to define the notion of stability for the system. Here, too, the mathematical definition needed can be given within the constructibility theory.

What is the point of giving such a mathematical description of the physical system? Geroch puts it as follows: "[T]opology is just a few definitions, a few constructions, and a few theorems which often happen, for some reason, to provide a convenient and appropriate framework for the description of the way nature actually does behave" (Physics, p. 186). It can be seen that the constructibility theory can serve this function of providing a framework for the description of the way nature actually behaves.

In view of the points made above, let us reconsider Shapiro's idea that Structuralism provides a more holistic view of science and mathematics than other philosophies of mathematics, in so far as it views scientific theories as incorporating mathematical structures. Let us allow that mathematical structures are frequently incorporated into scientific theories, as was done in the previous example from dynamics. The question I would like to raise is this: Why should this fact about scientific theories support Structuralism? In particular, why cannot the author's view being developed in this work accommodate this fact? One can surely develop, using set theory, a sophisticated and complex theory of mathematical structures; so any mathematical structure the physicist is likely to use can be treated within the framework of the mathematical theory of Chapter 6. I see no reason why the physicist could not make use of the standard structural definitions and theorems in constructing her descriptions of physical systems, without committing herself to the metaphysical baggage postulated by such Mathematical Structuralists as Resnik. Why do we need the Structuralist's principal claim (namely that the subject-matter of mathematics consists of structures) to adopt the "holistic view" that scientific theories incorporate mathematical structures?

Or consider another "advantage" to Structuralism cited by Shapiro: mathematical structures frequently share common sub-structures, and this accounts for the interconnections among the various branches of mathematics. Cannot the view of mathematics emerging in this work accommodate this fact? The constructibility theory can obviously be used to develop model theory: structures can be specified in pretty much the standard way. Thus, it would be a simple matter to show, by means of the constructibility theory, that the mathematical structures studied in the principal branches of mathematics frequently share common substructures.

5. An Evaluation of Resnik's Account of Structures

Let us now consider the plausibility and coherence of Resnik's radical ontological views about structures. There are several reasons for finding these views implausible.

1. The views, it would seem, cannot even be fully articulated without undermining their own truth. For the doctrine of the

nonsensicality of trans-structural identity asserts that objects from different structures cannot be said to be identical or not identical; but this very statement makes use of the notion of different structures— structures that are not identical—which, according to Resnik, is nonsense: structures cannot meaningfully be said to be not identical. It is not surprising that we find Resnik violating his own rules of meaningfulness time and time again, as when he says: "[W]e must evolve a comprehensive theory in which positions from different patterns can be dealt with in the same breath" (MaSoP:O, p. 538).

2. Resnik's views seem to lead to a rejection of a version of the principle (sometimes referred to as Leibniz's Principle) that if $X = Y$, then any open-sentence true of X is true of Y. For the number 2—an object in the natural number structure—is supposed to be a genuine object. So also is the set whose only member is the unit set containing the empty set. Now I can say of this set that it has a member; but according to Resnik I cannot say this of the number 2. But if I cannot say this, then presumably it is not true that the number 2 has a member; for if it were true, then surely it would be correct to say it. Yet according to Resnik, we cannot conclude that the number is a different object from the set. Schematically, we seem to have objects X and Y which are such that the open-sentence 'x has a member' is true of one but not the other; so by contrapositive of the above principle, it would seem to follow that X is different from Y. Resnik, however, maintains that it makes no sense to say X is different from Y. Surely all of this is not easy to accept.

I wish to make clear that I am not claiming that Resnik's position is logically impossible. However, I am claiming that it is implausible and counter-intuitive. For in flouting a principle that is equivalent (at least according to classical logic) to Leibniz's Principle, Resnik is violating a very intuitively plausible position. If X and Y are genuine objects (that truly exist) and if open-sentence F is true of X but not true of Y, it would seem that X would have to be a different object from Y. For if they were not different objects—in other words, if X were identical to Y—then surely any open-sentence true of X would be true of Y also. Resnik's view is certainly not easy to accept.

3. A final difficulty is discussed by Resnik himself. Resnik makes heavy use of the notion of congruence of structures. But how is congruence to be defined? Evidently, it would go something like the following: A is congruent with B if, and only if, there is a one-one correspondence between the objects in A and those in B such that . . .

But notice that we are here using a language in which we speak of the objects in both *A* and *B*. As Resnik notes: "This seems to require a universe which contains positions from different patterns" (MaSoP:O, p. 538). This seems to undermine Resnik's position, because in such a language, it would seem, that we would have to have the identity relation well-defined over the universe of this language.

Resnik makes several proposals for dealing with this problem, but in the end these proposals boil down to just two. The first is to substitute for "the motley of patterns dealt with by the various branches of mathematics and by [Resnik's] initial approach to patterns", one big pattern and "ask the mathematician to make do with it" (MaSoP:O, p. 539). But later, Resnik concludes that this proposal is unsatisfactory:

> [The] reduction to one "big" pattern would tend to push the relationships among patterns and their positions out of focus. And the discomforting interpattern questions cannot be avoided by the move to one pattern anyway, because we will still wonder how it relates to the old patterns or how it will relate to patterns discovered in the future (p. 542).

Resnik's second proposal for dealing with the above difficulty is to advocate a many-sorted logical language in which each universe of the language corresponds to the totality of objects of some structure talked about. In such a language, one can have infinitely many identity relations, one for each universe of discourse. One can thus avoid having any identity relation between the objects of different structures.

But will this do as a solution to the difficulty? Well, the following statement is surely true:

> Every structure with a substructure congruent to that of the natural numbers (omega, $+$, $\times$) has infinitely many substructures that are all different.

But one cannot say this in the language described above. In fact, one cannot quantify over structures at all. One cannot make general statements about structures. In particular, one cannot say 'Every structure is an object'—something that Resnik wants to assert. What is especially peculiar about this position is that, since it is maintained that structures are objects, the view seems to demand that it be permissible to quantify over these objects.

Resnik is a Platonist of sorts: he believes in the existence of abstract mathematical objects. But by characterizing these objects as mere positions in a structure and by adopting his extreme doctrine of the nonsensicality of trans-structural identity, he thought he could avoid the chief philosophical problems that have plagued the traditional Platonic views of mathematics. But as the above shows, he has merely exchanged one set of problems for another.

It should be pointed out that the author's constructibility theory is not plagued by the two particular philosophical difficulties that beset traditional Platonic views of the nature of mathematics and that Resnik attempted to resolve with his doctrine of the nonsensicality of trans-structural identity. How can we (who live in the physical world) have knowledge of the mathematical objects, supposedly talked about by the mathematician, which do not exist in the physical world? Since none of the theorems of the constructibility theory asserts the existence of non-physical abstract objects, knowledge of the truth of the assertions of this mathematical theory does not require that we perceive or interact causally with any non-physical objects referred to by the theory. The second problem, arising out of the fact that no mathematical theory can do more than determine its objects up to isomorphism, requires the traditional Platonist to maintain both that a given mathematical theory is about particular mathematical objects and also that it is impossible ever to pick out, by means of our mathematical descriptions, any particular mathematical objects. This difficulty is avoided in the constructibility theory because the mathematical theory is not held to make assertions about any particular mathematical objects.

It hardly needs pointing out that the author's point of view is not troubled by "Benacerraf's multiple reduction problem", which Resnik resolves by appeal to his doctrine of the nonsensicality of trans-structural identity. Since neither sets nor natural numbers are asserted to exist in the constructibility theory, no problem arises as to which sets are the natural numbers or which particular set is the number two.

8
Science without Numbers

1. Introduction to Field's Instrumentalism

LET us reconsider the reasoning that led Resnik to his paradoxical ontological views. Basically, he made use of the following two ideas: (1) The "logical form" of mathematical statements should be taken at face value; and (2) the semantics of the language of mathematics should be taken to be the same as that of the language of science, and hence should be regarded as standardly referential "in the manner of Tarski". Underlying Resnik's reasoning, however, is the tacit assumption that the theorems of mathematics must be true, for without this assumption nothing about the existence of mathematical objects can be inferred from (1) and (2). Now why should we accept such an assumption? One reason, which has a ring of plausibility to it, is that we have been taught to believe that mathematical theorems are true and we never seem to go wrong in so regarding them. But such a reason would hardly be convincing to one who does not already believe that mathematics is true. A more weighty reason can be derived from the fact that mathematics has been spectacularly successful in science. It can be claimed that the mathematical theories we have developed, such as number theory and analysis, are an indispensable part of contemporary science. Mathematics is used in formulating scientific hypotheses, analysing scientific data, and drawing conclusions from scientific theories, so to deny the truth of mathematics would seem to make a complete mystery of how scientists have achieved their successes. "How", it might be asked, "could a false theory guide us so well in obtaining correct scientific theories, true predictions, and successful applications of our scientific theories?"

But this sort of argument for the truth of mathematics does not impress Hartry Field, who believes that he has an equally good explanation for the crucial role of mathematics in science—one that does not presuppose the truth of mathematics. At the heart of Field's explanation is a certain "conservation theorem", which tells us, very roughly and only as a first approximation, that any nominalistic conclusions that can be drawn from a nominalistic theory with the aid of mathematics could also be drawn without this aid, just using logic. By appealing to this conservation theorem, Field defends a "fictionalist" and instrumentalist conception of mathematics: the systems of mathematics used by scientists do make reference to mathematical entities, but that is no reason for believing in the existence of such things, for we can see, he argues, that this use does not require that the assertions of mathematics be true. As a result, Field adopts the radical view that the theorems of mathematics are, for the most part, simply false. As Field analyses mathematics, the entities referred to and described by mathematicians should be regarded as useful fictions and the mathematical theories that purportedly describe these entities should be taken to be useful, but theoretically dispensable, devices for drawing conclusions—useful in science because they facilitate the drawing of nominalistic conclusions from nominalistic premises—conclusions that could be drawn by means of longer deductions without mathematics. That, in broad outlines, is the position Field takes in response to the sort of reason given above for thinking that mathematics must be true. Now to some details.

2. The Conservation Theorem

First of all, we need to have some idea of what "nominalistic theories" and "nominalistic assertions" are. In this section, all the theories and sentences I shall be considering will be theories and sentences of first-order logic. I restrict the discussion here to first-order theories, even though Field carried out much of his argumentation in his book in terms of second-order theories and also despite the fact that there has been much discussion in the literature on Field's book focusing on properties of second-order theories and second-order logic. I do this primarily for the sake of convenience and simplicity of exposition. Later, I shall consider Field's programme within the framework of both first- and second-order languages.

In this section, then, an *assertion* will be a sentence of a first-order language. So what is a *nominalistic* assertion? Field tells us: "A nominalistic assertion is an assertion whose variables are all explicitly restricted to nonmathematical entities" (Comments, p. 240); and the formal content of describing a sentence of a theory as nominalistic is simply that it does not overlap in non-logical vocabulary with the mathematical theory to be introduced (Science, n. 8). Hence, although Field motivates his nominalism by claiming that it avoids certain epistemological problems and perplexities arising from the apparent need to postulate the existence of abstract entities, in fact Field's nominalism turns out to be directed at avoiding a certain sort of abstract entity, namely mathematical objects.

Now suppose that N is a nominalistic theory, i.e. a first-order theory whose assertions are all nominalistic, and that ZFU is Zermelo–Fraenkel set theory with urelements (to be described below). Then, it can be proved, as a straightforward corollary of a standard logical theorem, that for any sentence s of the language of N,

if $N + \text{ZFU} \vdash s$, then $N \vdash s$

i.e. any nominalistic sentence s derivable in the combined theory $N + \text{ZFU}$ is derivable in N alone. It follows that the combined theory is a conservative extension of N. Thus, the mathematical theory ZFU is said to be *conservative over N*. The above theorem, however, is not quite the conservation theorem I described earlier as lying at the heart of Field's fictionalist position. N and ZFU are not connected by a common non-logical vocabulary, so it is hard to see how ZFU could be very useful to a scientist working with N. We need "bridge laws" connecting the nominalistic theory with the mathematical theory; more specifically, we need to be able to assert the existence of not only extensions of mathematical predicates, but also extensions of non-mathematical predicates constructed using the vocabulary of N. This can be done by both adding to ZFU an existential axiom stating that there is a set of all non-mathematical objects and also changing the subset axiom so as to include conditions on subsets in which non-logical constants from N may occur: the result will be a slight extension of ZFU, and the conservation theorem Field uses states that this extension is conservative over N.

Now for some details. Let ZF be Zermelo–Fraenkel set theory. Modify the standard Axiom of Extensionality to the following

$$((\text{E}z)(z \in x) \,\&\, (z)(z \in x \longleftrightarrow z \in y)) \rightarrow x = y$$

to allow for the existence of *urelements*, i.e. objects that are not sets. We thus get Zermelo–Fraenkel set theory with urelements (or ZFU for short). To the vocabulary of ZFU, add the predicate 'M', which we shall interpret as meaning 'is a mathematical object'. Then add three axioms to its assertions: one asserting that everything that has an element is a mathematical object; another asserting that there is something, namely the empty set, that has no elements but that is a mathematical object; and a third asserting that there is a set of all objects that are not mathematical objects. Call this new theory ZFU^1.

Let N be any consistent theory whose vocabulary does not overlap ZFU^1. (N is to be our "nominalistic theory".) In order to ensure that we have a "nominalistic theory" that makes assertions only about the non-mathematical realm, relativize all the quantifiers in N to the non-mathematical. Thus, for any sentence s of N, let s^* be the sentence obtained by relativizing all occurrences of quantifiers in s to '$-M$', e.g. if s is '$(x)Fx$', then s^* will be '$(x)(-Mx \rightarrow Fx)$'; if s is '$(\exists x)Fx$', then s^* will be '$(\exists x)(-Mx \& Fx)$'. Then, take N^* to be the theory obtained by transforming every assertion s of N to s^*. Finally, let ZFU^2 be the theory obtained from ZFU^1 by allowing the vocabulary of N to appear in the Comprehension Axiom.

We can now give a relatively precise statement of Field's Conservation Theorem:

For every sentence s of N, if $N^* + ZFU^2 \vdash s^*$, then $N^* \vdash s^*$

This theorem is proved in several different ways in the appendix to (Science, ch. 1). For example, one proof of this theorem consists in showing that from a particular sort of model of $N^* \cup \{-s^*\}$, the existence of which can be proved from the assumption that $N^* \cup \{-s^*\}$ is consistent, one can construct a model of $N^* + (ZFU^2 \cup \{-s^*\})$.

3. Field's Cardinality Theory

It will be helpful at this point to examine Field's use of the Conservation Theorem in analysing common types of reasoning involving finite cardinality. In what follows, I shall not bother with relativizing all the quantifiers in the way required to apply the Conservation Theorem, although it should be clear to the reader how this could be done. Now consider the following inference:

A1 There are twenty-one aardvarks.

A2 On each aardvark there are exactly three bugs.

A3 Each bug is on exactly one aardvark.

Therefore,

A4 There are exactly sixty-three bugs.

In the Fregean tradition, the premisses A1–A3, as well as the conclusion A4, can be analysed as saying:

B1 The cardinal number of the set of aardvarks is twenty-one.
B2 For every object x that is an aardvark the cardinal number of the set of bugs on x is three.
B3 For every object x that is a bug, the cardinal number of the set of all aardvarks y which is such that x is on y is one.
B4 The cardinal number of the set of bugs is sixty-three.

The English sentences B1–B4 can then be expressed in the language of ZFU^2 to obtain the formal sentences [B1]–[B4]; and since [B4] can be derived in ZFU^2 from [B1]–[B3], it is thought that the intuitive reasoning has been validated. But notice, the English sentence B1 and the formal sentence [B1] (as standardly interpreted) make reference to the *cardinal number of the set of aardvarks*—a mathematical object. So for the nominalist Field, B1 and [B1] cannot be true. But Field does not want to deny that the premisses A1–A3 might be true. So he has to find an acceptable nominalistic analysis of the premisses A1–A3 and validate the reasoning in some other way. This is done by first providing some recursive definitions that allow us to use, as abbreviations of ordinary first-order sentences, sentences containing special numerical quantifiers. The abbreviations follow the pattern:

C1 $(\exists : > 1 : x)Fx = (\exists x)Fx$.
C2 $(\exists : > k + 1 : x)Fx = (\exists x)(Fx \ \& \ (\exists : > k : y)(y \neq x \ \& \ Fy))$.
C3 $(\exists : k : x)Fx = (\exists : > k : x)Fx \ \& \ -(\exists : > k + 1 : x)Fx$.

The premisses A1–A3 can then be expressed as the following axioms of an ordinary first-order theory N, the non-logical vocabulary of which consists of $\{A, B, O\}$:

[A1] $(\exists : 21 : x)Ax$.
[A2] $(x)(Ax \rightarrow (\exists : 3 : y)(By \ \& \ yOx))$.
[A3] $(x)(Bx \rightarrow (\exists : 1 : y)(Ay \ \& \ xOy))$.

Adding the vocabulary of N to that of ZFU and making the changes in axioms described above, we obtain the appropriate ZFU^2. Now in the combined theory $N + $ZFU^2, it can be proved that

[A1] is equivalent to [B1]
[A2] is equivalent to [B2]
[A3] is equivalent to [B3]

so [B$_i$] is called an "abstract counterpart" of [A$_i$] in ZFU^2 (i = 1, 2, 3)—abstract, because [B$_i$] makes reference to mathematical objects. Similarly, the conclusion [A4] is analysed in N as

[A4] $(\exists{:}63{:}x)Bx$

and it can be proved in N + ZFU^2 that

[A4] is equivalent to [B4].

Since it can be proved in ZFU^2, in ways given in standard text-books, that for example $3 \times 21 = 63$, it is a simple matter to derive [B4] in ZFU^2 from [B1]–[B3]. Thus, making use of the equivalences mentioned above, it can be shown that N + ZFU^2 ⊢ [A4]. Hence, by the conservation theorem, we have N ⊢ [A4]. In other words, Field can make use of standard mathematical analyses of cardinality in validating the intuitive reasoning, but he doesn't have to assume the truth of mathematics to do that—he just needs to make use of the conservatism of mathematics.

Field's basic strategy for dealing with reasoning involving finite cardinality is clear: ordinary cardinality statements are treated nominalistically by means of numerical quantifiers and formulated as sentences in a first-order nominalistic theory; the more traditional analyses of such statements, involving reference to mathematical objects such as sets, are regarded as abstract counterparts; and the deductions made in the combined nominalistic and mathematical theory would proceed from nominalistic premises to abstract counterparts, then (in the traditional manner within the mathematical theory) to the abstract counterpart of the nominalistic conclusion, and finally to the nominalistic conclusion itself. The nominalistic conclusion obtained in the combined theory may be relied upon because, as the conservation theorem shows, any nominalistic conclusion drawn in the combined theory from the nominalistic premises is derivable directly from those premises within the nominalistic theory. The mathematical theory can thus be regarded merely as a device for facilitating the deductions of nominalistic conclusions from nominalistic premises.

4. Field's Nominalistic Physics

Of course, the conservation theorem does not, by itself, show that mathematics is dispensable in principle in science, even assuming for the sake of argument that all scientific theories can be adequately expressed as formalized theories of quantificational logic. For contemporary scientific theories are not nominalistic by any stretch of the imagination: even supposing that such theories could be adequately reformulated in some logical quantificational language, practically all of the fundamental theories in physics make explicit reference to mathematical objects, such as numbers, functions, operations, fields, and spaces. Hence, Field undertakes a programme of producing first- and second-order nominalistic versions of physical theories. In fact, however, Field concerns himself almost exclusively with Newtonian gravitational theory, suggesting that the methods utilized in obtaining a nominalistic version of that theory could be carried over to most (if not all) other physical theories.

To see how Field hoped to "nominalize" physics, let us imagine that PNP is a Platonic version of Newtonian particle physics formalized as a second-order theory—Platonic in so far as numbers, functions, etc. are referred to in the theory. Field attempted to devise a version of PNP which would have the following features:

(*a*) It would be nominalistic;

(*b*) any nominalistic sentence *s** of the theory that could be deduced from the Platonic theory PNP could also be obtained from this theory with the aid of *mathematics*, that is, the appropriate version of ZFU described in the previous section.

Calling Field's nominalistic version of physics 'NNP', and appealing to the Conservation Theorem, we can then assert that PNP is a conservative extension of NNP; for any nominalistic conclusion that could be drawn in PNP could also be drawn in NNP. Field believed that, in this way, he would show that "even a nominalist is free to use the platonistic formulations of physics in drawing nominalistically-statable conclusions; . . . these conclusions would follow from the nominalistic physics alone" (Realism, p. 20). Thus, for Field, "platonistic formulations of physical theories are simply conservative extensions of underlying nominalistic formulations" (p. 20).

But how did Field attempt to construct such nominalistic versions of physics? Field's strategy is modelled after Hilbert's axiomatization of Euclidean Geometry. Instead of having terms in the theory for mathematical objects such as numbers and functions, the vocabulary of NNP includes comparative predicates. The first-order quantifiers of the theory range over space-time points, and the second-order quantifiers range over regions of these points. And it is argued that space-time points and regions are nominalistically acceptable—they are held to be physical objects on a par with molecules and atoms.

To see how Field proceeded, let us concentrate on the geometrical portion of the theory. As I mentioned above, there are no operation symbols in the vocabulary of the theory; so in particular, there is no operation symbol standing for the distance function. But there is a three-place predicate $xByz$, which is understood as meaning 'x is a point on a line segment with end-points y and z', and a four-place predicate $xyCzw$, which is understood as meaning 'the line segment with end-points x and y is congruent to the line segment with end-points z and w'. The axioms of the geometrical part of the theory are formulated in terms of these comparative predicates. Following Hilbert, we can then prove, bringing the mathematical theory, M, into play, a representation theorem that states that there is at least one function, d, which takes pairs of points into the real numbers in such a way that:

(a) for any points x, y, z, and w, $xyCzw$ if, and only if, $d(x, y) = d(z, w)$;

and

(b) for any points x, y and z, $yBxz$ if, and only if, $d(x, y) + d(y, z) = d(x, z)$.

This d-function allows us to obtain the abstract counterparts that correspond to the nominalistic statements of NNP; it thus allows us to go from the nominalistic theory into the Platonic mathematical theory and then return again to draw nominalistic conclusions from the theory.

5. Some Doubts about the Adequacy of Field's View

Field's view of mathematics is attractive for several reasons. The basic idea that mathematics is a sort of instrument for extracting what is implied by our scientific theories is not new: the Logical Positivists had proposed such a view many years ago. (Cf. Hempel's statement that "mathematics (as well as logic) has, so to speak, the function of a theoretical juice extractor" (Nature, p. 379).) But to make use of contemporary logical theory to make the Positivist's rather vague idea precise and clear certainly marks a significant advance. Furthermore, his strategy for dealing with the problem of existence in mathematics is both daring and clever: we can avoid having to believe in mathematical entities by simply maintaining that mathematical theorems are false. And there can be no doubt that Field has displayed considerable ingenuity in carrying out the above analyses of theoretical reasoning in physics. Yet, there are reasons for being doubtful about the adequacy of this striking view of mathematics. I shall list some of these reasons below. No one of these objections is decisive, but as a group they tend to undercut much of the attractiveness of Field's instrumentalism.

Let us examine more carefully the specific claims being made by Field. One of the central ideas of Field's instrumentalism is that mathematics is "a useful instrument for making deductions from the nominalistic system that is ultimately of interest; an instrument which yields no conclusions not obtainable without it, but which yields them more easily" (Science, p. 91). Mathematics, according to Field, is *conservative*; and this means, he tells us, that "any inference from nominalistic premises to a nominalistic conclusion that can be made with the help of mathematics could be made (usually more long-windedly) without it" (p. viii). But is the theory M (Field's formalized version of mathematics) conservative (in this sense) over NNP, the nominalistic second-order version of Newtonian particle physics Field has developed? Stewart Shapiro has shown that it cannot be. In particular, he proved in (Conserv) that there is a nominalistic sentence g^* of NNP which is such that $NNP + M \vdash g^*$, even though g^* is not provable in NNP alone.

First of all, it should be pointed out that Field does prove in (Science) a second-order version of the Conservation Theorem. But what is proved is that ZFU^2 is *semantically* conservative over the nominalistic theory N, that is,

For every sentence s of N if $N^* + ZFU^2 \vDash s^*$, then $N^* \vDash s^*$

where '$N \vDash s$' is to be understood as saying that s is a second-order consequence of N. Since second-order logic is not complete, we cannot infer from this theorem that ZFU^2 is deductively conservative over N.

To see how Shapiro proceeded, we need only consider the geometric portion of NNP. Using the two comparative predicates B and C, we can define the idea of *an infinite set of collinear equally spaced points* (i.e. the idea of an infinite set of points all of which lie on a single straight line and which are such that the distance between any two adjacent members of this set is equal to the distance between any other two adjacent members of this set). Thus, there will be a formula of NNP, $F(R,p)$, which expresses the idea:

> R is an infinite set of collinear equally spaced points containing p as an end-point.

Now one can think of R as the set of natural numbers and p as zero, since such a set will give us the basic structure of the natural numbers. Thus, in terms R and p, and using the comparative predicates of NNP described above, one can construct formulas of NNP that represent the successor relation, as well as addition and multiplication of natural number arithmetic. For example, '$z = x + y$' can be defined with the above-mentioned vocabulary by a formula that says the distance between p and x is equal to the distance between y and z. One can then prove in NNP the axioms of Peano arithmetic, formulated in terms of successor, plus, times, p and R as indicated. Then we can construct a Gödel sentence, say $q(R,p)$, formulated in the vocabulary of NNP in terms of R and p, and "saying" (using the standard Gödelian coding devices) that NNP is consistent. Letting q^* be the sentence

$$(R)\,(p)\,(F(R,p) \rightarrow q(R,p))$$

it is easy to see that q^* is not provable in NNP: since

$$(\exists R)\,(\exists p)\,(F(R,p))$$

is provable in NNP, if q^* were also provable in NNP, then one could prove the consistency of NNP in NNP—something that is impossible given the Second Incompleteness Theorem (assuming, of course, that NNP is consistent). However, q^* is provable in NNP + M, since the consistency of NNP is provable in M. So M is not deductively conservative over NNP as was suggested in (Science).

Field's response to this objection (in Comments) is striking. On the one hand, he gives up many of the claims that he made in (Science). For example, he gives up his doctrine that mathematics is a useful but theoretically dispensable instrument for carrying out deductions that could be carried out without the aid of mathematics. He admits that he was wrong to suggest that mathematics is deductively conservative. "What I should have said", he declares, "is that mathematics is useful because it is often easier to see that a nominalistic claim follows from a nominalistic theory plus mathematics than to see that it follows from the nominalistic theory alone" (Comments, p. 241). In other words, it is *semantic* conservativeness that is now claimed for mathematics—not deductive conservativeness. Thus, although Shapiro's reasoning has shown that M is not deductively conservative over NNP, it does not undermine the claim that M is semantically conservative over the nominalistic physics. This change in doctrine, however, detracts from the neat picture of the usefulness of mathematics that was presented originally in (Science). Mathematics, according to the new picture, is no longer seen to be a mere instrument for shortening deductions; and it could very well be an indispensable means of determining what does follow from nominalistic premisses.

Another problem with the new view is that second-order semantic consequence becomes a crucial notion. And it becomes reasonable to focus on the legitimacy of Field's use of the notion. After all, as this idea is originally explicated, one makes use of Platonic notions: it is done in terms of second-order models. So this puts the burden of explanation squarely upon Field: how can he, a nominalist, appeal to second-order consequence in explaining how the supposedly false theory ZFU^2 is useful? (I shall indicate shortly how Field attempts to take up this burden.)

Opting for semantic conservativeness was done in order to shield a significant portion of the claims Field made regarding the logical relationships between his second-order nominalistic version of Newtonian particle physics and the Platonic version PNP. But in the end, Field decides that, because of various doubts that he has about the legitimacy of using any form of second-order logic, the sort of view he had been defending does "not ultimately yield a satisfactory reply to the Quine–Putnam argument, and that attention to the case of first-order theories is required" (Comments, p. 255). Given that he is willing to admit that his second-order approach is not satisfactory, one wonders why he surrendered his attractive characterization of mathematics as an instrument for shortening deductions. For in the first-order case, semantic conservativeness and deductive conservativeness come to the same thing. There is no need to give up his claims about deductive conservativeness once he has decided to develop his nominalistic physics within the framework of first-order logic.

Field's principal doubt about taking the second-order approach is that second-order logic does not conform to the requirement that a logic ought not to make any existential assertions. No existential assertion, he believes, should be regarded as logically true; and since his (second-order) mereological logic contains existential assertions, he has doubts about this logic. I shall say more about Field's views about logic and metamathematics in the Appendix; but for now it is sufficient to note that Field gives up much of what he claimed and did in (Science) and tries to save what he can of his basic view of mathematics by concentrating on first-order versions of Newtonian particle physics.

In (Science), Field did sketch a first-order version of NNP. Let us call this theory 'NNP^1'. The variables of NNP^1 range over space-time regions. Field adds to the vocabulary of NNP various non-logical constants to compensate for the loss of second-order quantification. Thus, the vocabulary of NNP^1 contains a binary predicate S, which is to be understood as meaning 'is a subregion of', and a unary predicate P, which is to be understood as meaning 'is a point'. In this setting, points are taken to be minimal regions. So we have as an assertion of NNP^1

$$(x)(Px \longleftrightarrow (y)(ySx \rightarrow y = x)).$$

Using the predicate P, one can convert the first-order quantifiers of NNP into quantifiers of NNP^1 by relativizing the former to P.

Shapiro, however, has uncovered a serious weakness in Field's first-order theory: he shows that the representation theorem described earlier cannot be proven in the appropriate ZFU^2, as can be done in the second-order case. The reason again has to do with Gödel's Incompleteness Theorems. Roughly, it is shown that if the representation theorem could be proved, then a Gödelian sentence could be proven in NNP^1 + ZFU^2 that could not be proven in NNP^1 alone. But this would contradict the deductive conservatism of ZFU^2 over NNP^1.

Thus, within the framework of first-order logic, Field can no longer describe the use of mathematics in physics in the way he did in (Science) and (Realism): *he cannot claim that by means of the representation theorem one can pass from the nominalistic physics to the Platonic theory and then return, by the same means, to the nominalistic theory.* He cannot claim, as he did in (Realism), that the nominalist is free to use the Platonic physical theories in drawing nominalistic conclusions, on the grounds that these conclusions could be derived in his nominalistic versions of the theories with the help of a conservative mathematics.

Field argues instead that to reply to the Quine–Putnam argument one doesn't have to maintain that the nominalist is free to use the Platonic versions of physics: it is sufficient, he believes, to show that the nominalistic versions of physics he constructs are, in some suitable sense, *as good*. Thus, he argues that the nominalistic conclusions that can be drawn in Platonic versions of physics, but that are not forthcoming in his nominalistic versions, are "arcane" or "recherché". And he questions that there are any good scientific reasons for accepting such conclusions. For what reasons could there be for accepting these recherché statements? He suggests that there are only two kinds of reasons for preferring the Platonic theory that has these statements as theorems over the nominalistic theory that does not: (1) the Platonic theory may be simpler than the nominalistic; (2) there may be direct observational evidence support-ing the recherché theorems. Regarding the first possibility, Field suggests that although the Platonic theory may be simpler in formulation, the nominalistic theory is simpler in ontology and hence, as he sees it, the nominalistic physical theory actually comes out simpler than its Platonic counterpart (Comments, n. 17). As for the second, he asserts that "it is hard to imagine any situation" in which there is empirical support for these recherché theorems. Thus, his answer to his question "Is there any reason to believe in the truth of the excess consequences that one theory has but the other hasn't?" is a clear "NO". But I, on the other hand, would give a different answer.

Aren't there scientific reasons for accepting such sentences as $q*$? After all, do we not have reasons for affirming the consistency of NNP^1? Field says many things that make it clear that he is convinced of the consistency of NNP^1. Indeed, in (Is), Field implies that he knows that some first-order mathematical theory, which is much more complex, involved, and powerful than NNP^1, is consistent. So it would seem that he should also claim to know that NNP^1 is consistent. And if we grant that NNP^1 is consistent, then by the usual Gödelian reasoning, we can infer that $q*$ is true. Now Field has suggested that this is no reason for inferring the inadequacy of NNP^1 since the first-order Platonic version of Newtonian particle physics also cannot prove a number of Gödelian sentences that we can see to be true (see the final pages of (Science)). But I am not arguing that NNP^1 is inadequate because there are these true sentences that it cannot prove. I am simply noting that, contrary to what Field has argued, there are grounds for believing that such "recherché" nominalistic sentences as q^* are true—indeed scientific grounds. These are grounds that Field himself should recognize. So, contrary to what Field has suggested, there are scientific grounds for preferring the Platonic versions of physics over their nominalistic rivals in so far as the former makes a number of nominalistic assertions that the latter does not assert—assertions that we have grounds to believe are true.

But I have other doubts about the adequacy of Field's view of mathematics. These are less technical in nature, but they are, in my opinion, no less effective in undermining the basic position. Thus, a second sort of doubt about Field's view is that it is questionable that all contemporary theories of physics can be reformulated as "nominalistic theories" of the sort Field's instrumentalism requires; and I say this for two reasons: (1) as I suggested in Chapter 1, it is not clear that contemporary science can or should be formulated as a collection of first-order theories; (2) even if reasonable first-order versions of our scientific theories can be constructed, it is not clear that nominalistic versions of these theories can be produced in the way Field suggests. It is not obvious, for example, that one can construct a Fieldian version of Quantum Physics on the model of Field's work on Newtonian gravitational theory.[1] Interestingly, Field has recently admitted that he cannot, at present, really carry out for all of physics his nominalistic programme of reconstruction (Comments, p. 255). So there is room for doubt that Field's account of the role of mathematics in science is adequate for all physics.

A third problem with Field's view is the questionable ontology of Field's nominalistic physics. It is reasonable to regard space-time points and arbitrary regions of these space-time points as concrete physical objects on a par with electrons? Resnik has argued that the role of the space-time points in physical theory is much more similar to that of mathematical objects than to that of electrons: "In contrast to the case of electrons, forces or planets, no particular body of observable phenomena led to the introduction of space-time points" (How, p. 167). Furthermore, ordinary physical objects are thought to occupy regions of space during intervals of time. But a region of space points cannot be thought to occupy space. Presumably, a region of

---

[1] For some specific reasons for thinking that quantum mechanics would raise problems for Field, see Malament (Rev, pp. 533–4). Cf. however Burgess's method in (Synthetic).

space is something that can be occupied—not something that does the occupying. Besides, regions of space-time points seem to be so little different from sets of space-time points, that it becomes hard to see just what motivation there is to Field's nominalism.

This doubt can be accentuated by noting that, with only a slight increase in vocabulary over Field's NNP[1], one can obtain a theory of space-time points and regions that is quite similar to classical real number theory and analysis. (See Resnik's (Ontology), sect. 2; cf. Burgess's (Synthetic).) The basic idea is this: since a space-time point can be regarded as an ordered quadruple of real numbers and since regions of these space-time points can be regarded as sets of these quadruples, Field's nominalistic ontology seems to provide us with all the structure needed to construct a "nominalistic" version of classical analysis. Thus, one can develop a powerful version of analysis by making use of Field's ontology of space-time points and regions, replacing talk about real numbers and sets of real numbers by talk about space-time points and regions. Of course, many would see this as a *reductio* of the hypothesis that space-time points and regions of these points are genuine nominalistic entities.

A number of commentators have noted that the continuum hypothesis becomes, under Field's theory of regions, a *physical hypothesis*, that is a hypothesis that is either true or false depending on what the physical facts are. This is because the hypothesis can be formulated as a hypothesis about regions of space-time points. (See, for example, Resnik (Ontology), p. 198.) The fact that this powerful set-theoretical hypothesis, which is undecidable from the standard axioms of set theory, is decidable in Field's "nominalistic" ontology may cause one to wonder at Field's claim that his ontology is genuinely "nominalistic". Certainly, this aspect of Field's position has engendered much sceptical comment (especially by Resnik in (How)), and needs to be defended much more convincingly than has been done thus far.

A fourth reason for being dissatisfied with Field's nominalism is to be found in the view of cardinality presented in (Science). The question is: How are everyday statements of cardinality to be understood? After all, even the nominalist will want to regard many ordinary cardinality statements to be true. Now Field suggests that such statements are to be understood in terms of inductively defined numerical quantifiers (above, Section 3). But this method will only work for a restricted class of statements involving finite cardinality. It is not clear, from what Field has provided us in (Science), how he would analyse the truth-conditions of even such well-known examples as

There are more dogs than cats.

In (Science), ch. 9, sect. II, Field introduces a "fewer than" quantifier to deal with such examples and claims that in his formal version of Newtonian gravitational theory, this quantifier can be defined in terms of a "there exist finitely many" quantifier if a new predicate is added to the vocabulary. This latter quantifier, he suggests, can be taken to be a primitive of his theory; but he tells us that a truth theory of this latter quantifier would "of course have to use the notion of finiteness" (p. 125, n. 59). An analysis of cardinality that does not presuppose the notion of *finiteness* (clearly, a cardinality notion) in this way would be preferable.

Another possibility is for Field to use the Quine–Goodman explication of 'fewer than' in terms of "bits" (that is, nominalistic parts),[2] but I suspect that he would be quite reluctant to resort to such a cumbersome analysis. Besides, the Quine–Goodman approach does not carry over to all cases of statements of the form 'There are more Fs than Gs', "regardless of how the individuals concerned overlap one another".[3] So the Fregean approach taken in this work would have definite advantages.

I wonder, too, how Field would understand statements of the form

The number of Fs is a prime multiple of the number of Gs

or

The number of Fs is an exact fraction of the number of Gs.

It is not hard to imagine a situation in which we might want to accept such statements; but it is unclear how Field's battery of cardinality quantifiers could be used to provide the truth-conditions for such statements.

Consider now this fifth reason for questioning the overall adequacy of Field's intrumentalism. To have a truly convincing instrumentalistic account of mathematics, Field needs to explain more than just why mathematics has been useful in science. He needs to explain both why mathematics has been so useful in *logic* and also how, in his own

[2] See Quine and Goodman (Steps, pp. 180–2).
[3] Ibid., p. 181.

theorizing about the applicability of mathematics in science, he can make use of the standard theorems of classical mathematical logic— theorems, such as the completeness of first-order logic and the Gödel Incompleteness Theorems, not only proved using mathematics, but which even seem to presuppose mathematics in the very definitions of key terms occurring in the statements of the theorems. Field has attempted in (Is) to remedy this defect in his position, by setting forth a kind of nominalistic justification for his use of model theory and proof theory. But there are many serious problems with his argumentation, as I shall point out in the Appendix.

A sixth reason for having doubts about Field's position is closely related to the above points. Field makes use of mathematics in his own theory of why mathematics is useful in physics: *the conservation theorem* described above, for example, *is proved using classical set theory*. How can Field make use of set theory in this way? He explains:

It may be thought that there was something wrong about using platonistic methods of proof in an argument for nominalism. But there is really little difficulty here: if I am successful in proving *platonistically* that abstract entities are not needed for ordinary inferences about the physical world or for science, then anyone who wants to *argue* for platonism will be unable to rely on the Quinian argument that the existence of abstract entities is an indispensable assumption. The upshot then . . . is that platonism is left in an unstable position: It entails its own unjustifiability (Science, p. 6).

Such a reply would have some force if the conservation theorem were only used to refute the Platonist's argument. But in fact, Field uses the theorem to provide himself with an *explanation*: he uses it to explain to himself and his fellow nominalists why mathematics is useful in science, writing: "According to the form of nominalism I accept, one should not junk a platonistic explanation of a phenomenon unless there is a satisfactory nominalistic explanation to take its place" (Is, p. 544). This is why he puts forward an explanation of why Platonic mathematics is useful in science. But how can he believe his own explanation, if the theory he appeals to in justifying a crucial part of his explanation is believed by him to be false? To see this more clearly, imagine that Field is trying to explain to a fellow nominalist how mathematics is useful in science. Can he appeal to a theorem proved in a theory that this nominalist rejects? The above reply, then, doesn't explain how the nominalist can rely on his own set-theoretical reasoning, when it is not just a question of using set theory hypothetically to undermine an argument for Platonism.[4]

It may be replied that there are versions of the conservation theorem that do not require the sort of model-theoretic reasoning described earlier (cf. the proof-theoretic versions sketched in the appendix to ch. 1 of (Science)). Would it not be legitimate for Field to appeal to these versions? The trouble is that even the proof-theoretic versions make use of some mathematics. It needs to be kept in mind that it is not just set theory that is Platonic for Field; even elementary number theory asserts the existence of mathematical objects and is regarded as false by Field. Thus, Field asserts that "there is no reason to regard any part of mathematics as true" (Science, p. 1 of the preface). Furthermore, even the use of standard proof theory raises a problem for him, because (as he himself put it) "proof-theoretic concepts are defined in terms of mathematical entities, with the result that proof-theoretic reasoning becomes reasoning about mathematical entities" (Is, p. 533).

Field's recently published paper (Is) contains the beginnings of a reply to the above objection. But this reply is too complex to be given a brief analysis here. I shall provide my evaluation of Field's implied reply in the Appendix.

My seventh reason for doubting the adequacy of Field's view of mathematics is the least clear-cut of these reasons, but I believe it does point to something fundamental about the nature of mathematics, and perhaps accounts for a vague feeling I have had that the purely instrumentalistic view of mathematics, although attractive in many ways, does not, as it stands, provide an adequate account of how mathematics is in fact applied in science and everyday life. I shall briefly sketch one kind of reason for thinking that mathematics is more than an instrument for making deductions. Consider the following argument:

Mathematics has been developed over the years as a collection of informal systems of truths: it has been regarded as expressing truths, and mathematicians and scientists have operated on the hypothesis that the theorems of mathematics are, for the most part, true, constructing both mathematical and scientific theories on

---

[4] This objection was made earlier in my (Simple). See Resnik (Review) for a similar objection.

that hypothesis. The history of mathematics and science has, in a loose but convincing way, confirmed this hypothesis.

Hilary Putnam once adopted a position that is close to the above. Putnam argued in (MMM, p. xiii) that science, including mathematics, attempts to tell a "unified story" which gives at least an approximation to the truth. And he went on to argue, quite forcefully, that the success in science of a mathematical theory or postulate, as for example that there is a one-to-one order preserving correspondence between the points on a line and the real numbers, provides us with grounds for believing in the truth of the theory or postulate in question (What, pp. 65–6). Underlying this argument is the use of a sort of "hypothetico-deductive" method of testing and confirming mathematical hypotheses. Thus, Putnam asserted that "the real justification of the [differential and integral] calculus is its success— its success in mathematics, and its success in physical science." More generally, Putnam argued, "the hypothesis that classical mathematics is largely true accounts for the success of the physical applications of classical mathematics" (What, p. 75), the suggestion being that without this hypothesis, the success of science would be inexplicable.

Now it may seem that Field has undermined these considerations. After all, hasn't he shown that mathematics can be "applied" in science without being true? And hasn't he provided an account of science that requires only a conservative mathematics? And doesn't this show the valuelessness of the above reasoning? Actually it does not. For Field's account of mathematical and scientific reasoning is directed at another argument (also given by Putnam) for belief in the truth of mathematics. This argument, although closely related to the above, is quite distinct from it. The argument that Field is attacking is just lurking beneath the surface of Quine's argument discussed in Chapter 1, and was developed by Putnam as follows: Our present-day scientific theories require us to believe in the existence of such mathematical objects as numbers and functions, because the rejection of the existence of mathematical objects would commit us to giving up not only much of mathematics, but physics as well. Even such a physical law as Newton's Law of Universal Gravitation (which is held to be approximately true) requires acceptance of mathematical objects. Putnam describes this argument in a later work with the words:

[T]he Law of Universal Gravitation makes an objective statement about bodies—not just about sense data or meter readings. What is the statement? It is just that bodies behave in such a way that the quotient of two numbers *associated* with the bodies is equal to a third number *associated* with the bodies. But how can such a statement have any objective content at all if numbers and 'associations' (i.e. functions) are like mere fictions? (What, p. 74).

Thus, Putnam took the position that to assert the truth of such a physical law as Newton's Law of Universal Gravitation, one would have to quantify over such entities as "*functions from* masses, distances, etc. *to* real numbers". He concluded that "a reasonable interpretation of the *application* of mathematics to the physical world *requires* a realistic interpretation of mathematics" (MMM, p. 74).

As I mentioned above, this argument is not Quine's. But it is sufficiently close to Quine's argument to deserve Quine's name somewhere in its title. Let's call it Putnam's *neo-Quinian argument*.[5] Basically, it is in response to this neo-Quinian argument that Field attempts to show how present-day scientific theories can be reconstructed as theories that do not presuppose the truth of mathematics, and which require mathematics only as an instrument for making deductions. I wish to emphasize that Putnam's neo-Quinian argument and Field's response to it are concerned with *present-day* science. However, the reasoning given above for thinking that mathematics enunciates truths is not based on features of present-day science. Rather, it is based on the nature of mathematical and scientific thinking throughout history. To see this more clearly, let us consider briefly an objection to Field's instrumentalism that bears on the above point.

Jonathan Lear has argued that the conservativeness of such mathematical theories as ZFU does not adequately explain the usefulness of the theory in science; for one can construct indefinitely many theories that are conservative over, say, NNP, but which are not of any use at all. He claims: "It is precisely because mathematics is so richly applicable to the physical world that we are inclined to believe that it is not merely one more consistent theory that behaves

---

[5] Field seems to think that Putnam's argument for Platonism just is Quine's argument, writing: "The most thorough presentation of the Quinian argument is actually not by Quine but by Hilary Putnam" (Science, n. 4). Such an identification can lead to trouble. It needs to be kept in mind that Quine's argument is inextricably bound to doctrines about the nature of language, science, ontological commitment, and ontology.

conservatively with respect to science, but that it is true" (Aristotle's, pp. 188–9). The implication here is that only by assuming the truth of the mathematical theory can we explain why it is so useful in drawing empirical conclusions from empirical premisses and why it is "so richly applicable" to the physical world.

Lear's reasoning, however, is not entirely clear. What precisely needs to be explained? His argument proceeds from the premiss that one can construct indefinitely many conservative theories that are of no use at all in science. Let us grant that. But, equally, one can construct indefinitely many true theories that are of no use at all in science. Should we not conclude then that truth no more explains usefulness than conservativeness does? Such considerations bring out the need to specify more clearly just what needs to be explained and what may be presupposed in giving the required explanation. Lear's demand that an explanation be given of why mathematics is "so richly applicable" is similar to the demand made by realists, in the course of criticizing empiricists in the philosophy of science, that one explain why the scientific theories we have are so useful. Such demands can easily become unreasonable. After all, one doesn't want to require, as a condition of adequacy of any philosophical view, that it explain why the world is the way it is. We cannot require Field to explain why, of all the possible worlds, the actual world is one in which ZFU turns out to be useful.

Supposing, however, that it is reasonable to demand an explanation of why mathematics is so applicable, Field might offer a kind of Darwinian explanation of this. The conservative theories we now accept, he might argue, are the results of many centuries of selection: only the conservative theories that have proven to be useful have been retained. So it is not surprising that the ones we now have are useful.[6]

But if Field were to attempt to explain why our mathematical theories are so useful in science in the above Darwinian way, Lear could raise several objections to the explanation. First of all, mathematicians do not in fact discard mathematical theories just

---

[6] An explanation of this sort was proposed by Bas van Fraassen to explain, from an empiricist's standpoint, the usefulness of our present scientific theories: "I claim that the success of current scientific theories is no miracle. . . . For any scientific theory is born into a life of fierce competition, a jungle red in tooth and claw. Only the successful theories survive—the ones which *in fact* latch on to actual regularities in nature" (Image, p. 40). He also proposed such a Darwinian explanation of the usefulness of mathematics in a paper entitled "On the Reality of Mathematical Entities", which he delivered at the *Levels of Reality* symposium held in Florence in Sept. 1978.

because they are not *useful in science*.[7] So the picture of mathematics presented by the Darwinian explanation—one in which large numbers of conservative mathematical theories are produced (at random?) and then reduced by a process that selects only the scientifically useful ones—can be challenged. To claim that there is a process of selection for conservative mathematical theories that are useful in science would be controversial at best. Second, I shall argue, an examination of the history of mathematics will show that few mathematical theories were conservative instruments of deduction of the sort that Field's book describes. So, if I am right, there was in fact no rich flow of conservative theories from which the useful ones could have been selected.

Consider an example from seventeenth-century mathematics. Newton's physics was certainly not a nominalistic theory of the sort Field constructs. It was not even a first-order theory: it was formulated in a natural language. The fact that the physical theory Newton was developing at this time was not expressed in a formal language is significant; for *there is no obvious way of distinguishing the vocabulary of the physical theory from that of the mathematical theory.* Thus, it is hard to see how one can get the Fieldian sort of analysis of the usefulness of mathematics even started: how is one to apply the conservation theorem unless one can separate the vocabulary of the physical theory from that of the mathematical theory? But more importantly, Newton's physics was not a science without numbers: Newtonian mathematics entered into the very formulation and construction of the new physics. Clearly, the mathematics Newton used in his science was not operating as a mere instrument for deducing nominalistic conclusions from nominalistic premises. Indeed, Newton's mathematical reasoning does not even appear to be formalizable as the sort of first-order logical instrument Field describes in his book. The reasoning that went into the development of Newton's infinitesimal calculus was notoriously imprecise and unrigorous. Many of Bishop Berkeley's criticisms of Newton's mathematical thinking were well taken.[8] And Newton was certainly aware of unclarities in his own method of infinitesimals. Terms such

---

[7] The question of why certain mathematical theories are discarded is a complex one. See H. J. M. Bos (Equations) for one analysis of why the area of mathematics called the "construction of equations" fell into oblivion. Cf. Charles Fisher (Death) for a quite different sort of explanation of why the Theory of Invariants disappeared.

[8] See J. Grabiner (Cauchy, pp. 33–4), for some details of his criticisms.

as 'infinitesimal' and 'infinite sum' were not clearly defined, clearly explained, or even clearly understood. Reasoning was sometimes used that could not be recast as valid arguments that conformed to any precise rules of proof that were acceptable to the mathematical community of that time, and seemed to depend on the reasoner's "feel" for the subject-matter. There were even times at which Newton seems to have made use of "empirical methods" to help convince himself of the truth of mathematical hypotheses. For example, the generalization of the binomial theorem,

$$(a+z)^b = 1 + bz + [b(b-1)/2!]z^2 + \ldots$$

seems to have been accepted, not as a result of a proof, but rather as a result of extrapolation and analogical reasoning from the number-theoretic cases, coupled with the verification of the hypothesis for special cases (see Kitcher (Nature), pp. 233–4). And as is well known, Newton frequently appealed to geometric intuitions and to intuitions of movements in space to justify steps in his reasoning. Indeed, François de Gandt went so far as to say:

On pourrait dire, en forçant à peine: ce n'est pas la dérivée qui a permis la définition de la vitesse, mais le contraire. Dans un grand nombre de textes [du XVIIe siècle], la vitesse instantanée est une notion considérée comme admise, et qui sert de base aux raisonnements infinitésimaux. L'example de Newton est très net: son calcul des "fluxions" est une comparaison entre des vitesses de variation (Réalité, p. 167).

De Gandt concludes:

Le géomètre de XVIIe siècle se considère comme un ouvrier dans le grand royaume des courbes, mais il se préoccupe rarement de délimiter strictement les hypothèses de départ et les constructions admises. Les raisons de cette déshérence sont multiples: d'abord, les géomètres ont mis au point des procédés qui "marchent", mais qui restent injustifiables en termes euclidiens; d'autre part, et plus largement, les hommes de cette époque sont persuadés que le recours à la lumière naturelle rend superflue toute discussion sur les principes (pp. 190–1).

In short, Newton's mathematics was so intertwined with his physics that no separation of the scientific theory into two separate first-order theories with separate vocabularies, in the manner of Field, seems at all feasible.[9]

[9] Cf. "Le XVIIe siècle a vu naître en même temps le calcul infinitésimal et la science du mouvement, à peu près entre 1610 et 1690. Les deux directions de recherche sont inséparables" (De Gandt (Réalité), p. 167).

Summarizing the above points, I have argued that even if it is granted that Field has explained how certain sorts of false mathematical theories can be useful, this gives us no reason for thinking that the mathematical theories scientists have used in the past were useful in the way described in Field's explanation; for it is clear that the actual mathematical theories scientists used in the past were not the sorts of instruments for deriving nominalistic conclusions from nominalistic theories that Field's explanations dealt with. Furthermore, Field cannot respond to the kind of demand discussed earlier for an explanation of why mathematics is so richly applicable in science by giving the Darwinian explanation championed by van Fraassen: scientists and mathematicians have not been winnowing a large hopper of conservative deductive instruments, selecting those that turn out to be useful in science. Such a picture simply does not represent what occurred in past centuries.

These considerations are also relevant to the argument in favour of regarding mathematics as a body of truths. Thus, Newton's method of proceeding in mathematics makes sense from the perspective of one who regards the theorems of mathematics as true. From such a perspective, it would be perfectly natural to formulate one's scientific laws and principles using mathematical notions and mathematical theorems. It would be a different matter of course if one believed that mathematical theorems were false or meaningless. Furthermore, the fact that fundamental notions of the mathematics used were unclear, that the reasoning employed was sometimes vague, intuitive, or unformalizable, that hypotheses were accepted, if only provisionally, on the bases of verification by small numbers of special cases, can all be understood from this perspective: all this is analogous to what is done in the pursuit of scientific truths and seems perfectly reasonable when regarded as a means of arriving at mathematical truths. Thus, throughout history, scientists have made use of notions that were unclear, improperly defined, or poorly grasped, by present-day logical standards, and yet played significant roles in the development of scientific theories that we now regard as true and important. Ideas such as that of solid, liquid, gas, gene, living organism, element, and chance variation, are examples. And scientists frequently make use of reasoning that does not proceed according to precise rules of inference, but that instead relies on the scientist's "feel" for the subject-matter or on his/her intuition of how to proceed, based on various paradigmatic analyses already encountered. This is something that Thomas Kuhn has emphasized in his studies of both past and present scientific reasoning (as for example in (Revolutions), especially ch. 5). And, of course, the provisional acceptance of a scientific hypothesis, based on the verification of a number of special cases, is not at all unusual in science.

Nor is the above example of Newtonian mathematics and science an isolated case. What was said above describes, in varying degrees, what mathematics was like throughout history. Consider the following example, which philosophers of mathematics are fond of using.[10] The eighteenth-century mathematician L. Euler determined the value of the series $1/n^2$ by a clever form of reasoning that certainly would not be considered a *proof*, by present-day standards. Basically, he divided the power series expansion of $\sin x$ by x to obtain the equation

[*] $(\sin x)/x = 1 - x^2/3! + x^4/5! - x^6/7! + \cdots$

The right side of the equation can thus be regarded as an infinite polynomial having as roots those values of x for which $(\sin x)/x = 0$, i.e. $x = +\text{pi}, +2\text{pi}, +3\text{pi}, \ldots$. Thus, he was able to represent $(\sin x)/x$ as the infinite product

$(1 - x^2/\text{pi}^2)(1 - x^2/4\text{pi}^2)(1 - x^2/9\text{pi}^2) \ldots$

Expanding this product, he obtained as coefficients of x^2

$-(1/\text{pi}^2 + 1/4\text{pi}^2 + 1/9\text{pi}^2 + \cdots)$.

He then identified this coefficient with the coefficients of x^2 obtained above in [*], i.e. $-1/3!$. Finally, multiplying both sides of this equation by $-\text{pi}^2$, he obtained as value of $1 + 1/4 + 1/9 + \cdots$ the simple quotient $\text{pi}^2/6$.

Of course, this reasoning is not at all rigorous. But Euler gathered additional evidence for his result that made his conclusion very convincing. For example, he calculated partial sums of the above infinite series and verified that they approached $\text{pi}^2/6$.[11] That such modes of reasoning are used in mathematics is perfectly reasonable if

---

[10] This example is discussed in masterful fashion by Polya in (Induction, sect. 2.6). It is also presented in Steiner (Mathematical, pp. 103–6), Putnam (What, pp. 67–8), and Kitcher (Nature, pp. 241–3). And as early as 1963 in (Discovery), I cited Polya's work as providing striking analyses of mathematical reasoning which supported a veridicalist view of mathematics.

[11] See Steiner (Mathematical, pp. 105–6) for a discussion of some additional evidence that Euler provided.

one regards mathematicians as attempting to determine truths. Recall that, although conservativeness is a property of a deductive theory as a whole, truth is something that can be attributed to individual sentences (or propositions). The mathematical methods and results that Euler used to determine the values of infinite series cannot be accurately formalized so as to yield a coherent and consistent system. Euler was aware of the fact that the methods he used sometimes led to anomalous results and some of the mathematical statements he accepted (such as '$1/4 = 1 - 2 + 3 - 4 + \cdots$') have no clear sense by present-day standards. But methods that may be quite questionable when applied generally may be reasonable and secure in individual cases, when it is the truth of the individual statement that is at issue—especially if the result of applying the method can be independently tested and checked.

Another striking feature of the development of mathematics is illustrated by the above example: the opening up of a new area of mathematics has frequently led to discoveries that were verified by an older, more established area. Thus, the use of infinite series, limits, trigonometric functions, integrals, and other notions of analysis, even when only poorly articulated and sometimes only vaguely understood, were instrumental in arriving at and justifying mathematical theorems that were subsequently verified by number-theoretic methods. This is just the sort of development one would expect on the hypothesis that mathematics is a system of truths and mathematicians are attempting to arrive at truths. The success of this way of proceeding in mathematics has been remarkable. It is hard to imagine how so much could have been achieved in science without the kind of mathematics we have had.

So far, I have only appealed to considerations of a very general sort to support my contention that mathematics is a body of truths. In the following, I shall emphasize specific theorems and specific mathematical results. Consider the *fundamental theorem of integral calculus*. It says that, for any real-valued function of a real variable, f(x), that is continuous in the interval (a, b), if F(x) is an indefinite integral of f(x), then

$$\int_a^b f(x)\, dx = F(b) - F(a).$$

Suppose, for reasons of simplicity of exposition, that f(x) is positive for all x in (a, b). Then the definite integral $\int_a^b f(x)\, dx$ gives the area under the curve determined by $f(x)$ that is bounded by a, b, and the x-axis. For an enormous number of examples, one can plot the curve on engineering paper and estimate this area by simply counting the whole squares in this area and estimating the partial squares. Hence, for all such examples as a student of calculus is apt to be given, one can verify that $F(b) - F(a)$ does indeed give the value of $\int_a^b f(x)\, dx$—at least to the degree that one can approximate the areas in question. Surely, this provides us with some reason for thinking that there is at least some truth to what the theorem says.

Additional support for this belief can be obtained by developing classical analysis within the framework of the mathematical theory described in Chapter 6. In this setting, it would be a simple matter to prove what amounts to the fundamental theorem, and, without appealing to abstract mathematical objects, we could gain an insight into why calculating closer and closer approximations of the area under the curve generated by $f(x)$ in (a, b) must approximate the value $F(a) - F(b)$. (I shall return to this idea in Chapter 12.)

The fundamental theorem is not an unusual case: there are countless theorems of this sort that, in effect, give us methods for calculating things that we can verify by methods and procedures that can hardly be questioned. Take Descartes' influential work on algebraic curves. Although this work was directed primarily at solving geometrical problems, the development of his methods led to the solution of algebraic equations by geometrical means. For example, he discovered a method of solving third- and fourth-degree equations through the intersection of a parabola and a circle (see Bos (Curves), p. 330). Many of Descartes' mathematical assertions could be checked by simple arithmetic calculations. Others could be checked by making elementary geometrical constructions. Can all these assertions be false?

Thus, to repeat the central idea of this argument against Field, mathematics has been regarded for hundreds of years as a developing body of truths; mathematicians and scientists have acted on this way of regarding mathematics: they have reasoned and constructed their theories with the tacit belief that the accepted statements and principles of mathematics are, for the most part, true. Many of these mathematical beliefs have been checked and rechecked countless times, and in countless ways, by both sophisticated and elementary methods. Furthermore, this way of proceeding has yielded remarkable results and has been tremendously successful. We thus have some strong reasons supporting the belief that mathematics is a body of truths.

The above considerations are by no means decisive. Still, in the absence of significant reasons for doubting this conviction of the veridicality of mathematics, it would seem reasonable to take mathematics as, for the most part, true. So it is worth asking if Field has provided us with any good reasons for his doctrine that mathematical theorems are false. Evidently, the only reason Field has for thinking that mathematical theorems are false is his belief that *one cannot grant the truth of mathematical theorems without committing oneself to an ontology of abstract mathematical objects* and all the epistemological difficulties that this brings with it. But can we not allow that mathematics, as a science, does contain much truth, without becoming thereby mathematical Platonists?

6. A Comparison with the Constructibility Theory

My position is that mathematicians do discover truths. However, I deny that these truths are about mathematical objects that transcend the physical world. Let us reconsider the six reasons given earlier for questioning Field's instrumentalistic view, to see how the constructibility theory fares when faced with such objections. My reasons for questioning the overall adequacy of Field's instrumentalism can be given as a series of doubts:

1. Doubts about Field's claim that there are no scientific grounds for preferring Platonic versions of Newtonian particle physics over his nominalistic first-order version.

2. Doubts about the expressibility of present-day science in the required nominalistic form Field's view requires.

3. Doubts about Field's "nominalistic" ontology of space-time points.

4. Doubts about the adequacy of Field's account of cardinality.

5. Doubts about the reasonableness of Field's logical theory, and in particular of his *acceptance* of model-theoretic logical theorems that have only been explained and proved using mathematics.

6. Doubts about the reasonableness of Field's appeal to the conservation theorem to explain the usefulness of mathematics.

7. Doubts about the reasonableness of regarding essentially all the theorems of mathematics as simply false.

Now the constructibility view of this work does not depend on defending Field's version of Newtonian particle physics, so the first doubt is not relevant to the constructibility theory. Similarly, the constructibility theory does not require a number-free, first-order reformulation of all of science, so the second of these doubts does not pose a serious problem. It is worth pointing out that the constructibility theory is not essentially tied to the first-order formulation I gave earlier. As the use of the constructibility theory is not essentially tied to a conservation theorem of first-order logic, there is no reason why the mathematical theory could not be given in ordinary English and applied to physical theories expressed in English. Nor does the view require an ontology of space-time points and regions of space-time points. Thus, the third doubt does not apply. Cardinality theory, both finite and infinite, can be developed in standard ways within the constructibility theory, as was sketched in Chapter 5, so the constructibility view does not fall prey to the fourth doubt. From the point of view of the constructibility theory, mathematics can be applied in logic in a relatively straightforward way, so doubt five poses no problem. Since the constructibility theory does not explain the usefulness of mathematics in science by means of the conservation theorem, the sixth doubt does not apply to it.

The seventh doubt needs to be treated in more detail since it raises some fundamental questions. What I say here is meant to provide only a preliminary response to this doubt—a fuller response will be given in Chapter 12. The constructibility theory is not a theory about how to analyse actual mathematics. I have not been claiming that the existential quantifier in ordinary mathematics should be treated as a constructibility quantifier. However, I have been arguing that the kind of mathematics studied in this work could provide an adequate mathematical framework for science. In other words, I have in effect been providing a response to what I have called Putnam's neo-Quinian argument—a response that implies that we do not have to believe in the existence of mathematical objects simply because present-day science seems to involve reference to such objects in the way it does. Clearly, this response does not require that the actual mathematics we now have be construed as a mere device for making deductions. It is perfectly compatible with the response I give to

Putnam's neo-Quinian argument that our actual mathematics enunciate statements that are for the most part true or approximately true—at least when these statements are construed in certain ways, say, in ways fashioned on the model of the constructibility theory. Thus, the view being set forth in this work is not committed to a position that is undermined in any direct fashion by the seventh doubt. However, more needs to be said in clarification of my position, and as I indicate above, a more complete picture of it will emerge in Chapter 12.

7. A Re-examination of Resnik's Reasoning

Given my previous argument that we have grounds for taking mathematics to be true, it may seem that I will run into difficulties with Resnik's argument that the "logical form" of mathematical theorems should be taken at face value and that the semantics of the language of mathematics should be construed as straightforwardly referential. It is time to return to Resnik's doctrine discussed earlier. I agree with Field in questioning the truth of mathematics, *taken at face value and regarded as straightforwardly referential*. But it is not obvious that "the truths" which our mathematical theories attempt to express should be understood in this referential way. Might not the truths underlying our mathematics be truths about constructibility?

In reply, Resnik makes use of Benacerraf's idea that the semantics of the language of mathematics should be taken to be the same as that of the language of science, and hence should be regarded as standardly referential "in the manner of Tarski". But one rather large assumption of this reasoning is that we already have a satisfactory semantic theory of the languages used in science and that this theory has been found to be the standard referential one. I do not find this assumption at all evident. It is simply not the case that the actual languages scientists use in stating and applying their theories have been shown to have the referential semantics of mathematical logic. After all, no one has been able to do for natural languages what Tarski did for the language of mathematical logic.

Some details may help to bring home my doubts on this question. Resnik's idea that the "logical form" of statements of mathematics and science should be taken at face value needs to be examined carefully. Is it obvious that we should analyse such statements in the standard Tarskian referential way (as is done in introductory logic courses)?

Before attempting to answer this last question, we should be clear about just what it is that we want in an analysis of such statements. In view of the argumentation of earlier sections, we can agree that the mathematical theorems we learn, for the most part, express truths. However, we should not assume that these mathematical theorems are straightforwardly and literally true: it should not be assumed that, by the mere linguistic analysis of mathematical language, one can determine what truths have been discovered by mathematicians. I wish to emphasize that, for my purposes, it is not at all important what the correct linguistic analysis of mathematical language is. After all, Field may be correct in maintaining that classical set theory, literally construed, is simply false. What is important for us to uncover with our analysis is a body of truths underlying the theory which our mathematical theorems can be taken to express or approximate—whether or not the theorems, literally construed, express a great deal that is not captured in the analysis. I shall discuss this point about what we should expect of an analysis of mathematics in more detail in Chapters 9 and 12. For now, it is enough to keep in mind that the kind of analysis that is under consideration here is not something to be entrusted to the linguist, but is a logical and methodological undertaking, involving theory construction and the analysis of ordinary and scientific reasoning.

The following statements were found in some scientific papers:

(a) There are difficulties with this hypothesis.
(b) There is a slight possibility that the sample was contaminated.
(c) The length of the rod was shortened by 0.6 cm.

Now should we treat (a) as having the logical form

$$(\exists x)(\exists y)(Dx \ \& \ Dy \ \& \ xPh \ \& \ yPh \ \& \ x \neq y)$$

so that we understand the scientist to be asserting the existence of strange "entities" called 'difficulties'? And does (b) assert the existence of things that are called 'slight possibilities'? And did the scientist who asserted (c) assume the existence of objects that are lengths? However one may answer these questions in the end, much hard evaluation and weighing of alternatives should be done before accepting Resnik's idea. One doubts that Resnik himself would both take all the statements of science "at face value" (in this sense) and also accept them as true.

But consider other problems with Resnik's reasoning. Scientists—especially those who work in fields of science dealing with humans (I have in mind such fields as psychology, anthropology, sociology, linguistics, and human biology)—frequently make assertions involving terms of psychological attitude. They assert such things as 'Mary hoped that her room-mate would stop drinking, but she realized that it was not likely.' Now does Resnik believe that we already have a satisfactory semantic analysis of such terms within the referential framework of mathematical logic? Surely not. He is, I am sure, aware of the vast literature on this topic which suggests otherwise. Furthermore, as Putnam has emphasized (above, Chapter 3), scientists also make use of modal notions in their theorizing, notions which have resisted analysis in the standardly referential way.[12] Hence, it would seem that Resnik is not basing his reasoning on a semantic analysis of the actual language scientists use.

I suspect that what underlies Resnik's beliefs about the language of science is Quine's doctrine that we ought to have a kind of official language of science for use when we are describing the ultimate structure of reality. As I pointed out in Chapter 1, it is advocated by Quine in (W and O) for example, that we construct a "canonical logical language" (essentially first-order predicate logic) into which we are to paraphrase our scientific beliefs whenever we attempt to state what we truly believe about the real world. The idea is not that actual scientific languages have the standardly referential structure of Tarskian semantics. Instead, we are to try to imagine, by extrapolating from present trends in science, simplifying and clarifying the overall conceptual apparatus in the process, the sort of scientific language we would get as a sort of ideal. In short, we are again dealing with the kind of make-believe science that I discussed in Chapter 1. To repeat a point I made earlier, Quine does not seriously undertake the task of formulating science as a first-order theory. So to suggest that we already know that a first-order reformulation of all of present-day science would be both feasible and preferable to the actual theories we in fact have is highly suspect.

---

[12] Interestingly, Resnik himself makes use of modalities and terms of psychological attitude in his own theorizing. For example, in (How, pp. 177–8), he writes: "Intention and function play a part too in our recognition of the thing as a template of the appropriate kind. . . . While some drawings on blueprint paper can represent things that it is logically possible to build but fail to represent possible buildings, others might fail to depict by simply making no sense at all."

It is noteworthy that in so far as one can generate reasons for thinking mathematics is a system of truths, based on the history of mathematics and science, as I did earlier, the various mathematical theories actually applied by scientists of the past were not, for the most part, the first-order versions of set theory that philosophers, such as Quine, have in mind when they describe the language of mathematics and science as "standardly referential".

The Quinian view of scientific theories is a throw-back to the "received view" of the Logical Positivists.[13] I find it ironic that Resnik—a Structuralist—would accept a view of scientific theories that is so close to the Received View. One would think that he would have found the semantical approach to scientific theories described in Chapter 7 much more attractive, since the semantical approach emphasizes structures rather than the sentences of some first-order language. There are, of course, many reasons for preferring the semantical approach over the syntactical view which are independent of one's views about the nature of mathematics. For example, commenting on the nature of evolutionary theory, Paul Thompson has claimed that "the most significant advantage of the semantic account is that it quite naturally corresponds to the ways in which biologists expound, employ, and explore the theory" (Perspective, p. 227). And he supports—at least in the case of evolutionary theory—van Fraassen's position that the semantic characterization of theories provides a more fruitful way to view foundational research in the sciences than is provided by the more familiar characterization of science as partially interpreted axiomatized theories (p. 229).[14] I should add that the semantical approach to scientific theories is not incompatible with the position of this work. The mathematical structures required by such an approach can be specified within the constructibility theory.

8. The Neo-Quinian Argument Reconsidered

Reconsider my response to the neo-Quinian argument described above in Section 5. I argued that the mathematics of present-day

---

[13] See Suppe (Structure) for a detailed discussion of the Received View.

[14] For a detailed discussion of the semantical approach to evolutionary theories, see E. Lloyd (Evolutionary).

science can be produced within the framework of Simple Type Theory and that this mathematics, when interpreted in terms of the constructibility quantifiers, does not make any existential assertions. So, in particular, one can use this system to assert the Law of Universal Gravitation without having to quantify over such mathematical objects as numbers and functions.

It has been suggested by Gottlieb in (Economy), p. 130, however, that my use of the constructibility quantifiers may contain some sort of "hidden commitment" to mathematical objects. How can I be sure that my own view does not presuppose mathematical objects? Perhaps I cannot be sure. But is that a reason for believing in mathematical objects? Can I be absolutely certain that all reported sightings of UFOs are explainable as illusions, hoaxes, etc.? Not at all. Yet, I don't conclude from this that we ought to believe in UFOs. To put the point another way, if one is to argue that we ought to believe in mathematical objects because we can't be sure that my constructive quantifiers do not carry some sort of hidden commitment to mathematical objects, then we have a very different sort of argument for belief in mathematical objects from the one we started out analysing. It is hard to believe that very many philosophers would consider such an argument very compelling.

It is clear that my attitude towards the Platonism–nominalism dispute is very different from Gottlieb's. Gottlieb begins his book (Economy) with the "intuition" that *abstract entities are mysterious and must be avoided at all cost* and that *appeal to such entities is especially pernicious in mathematics* (p. 11). My intuitions on these basic ontological questions are less definite. It is not even clear to me that the question 'Do any abstract entities exist?' has a definite (correct) answer. But when I read a philosopher's claim to have "scientific evidence" to support the belief that mathematical objects exist or read a philosopher's elaborate epistemological theory of how we obtain knowledge of mathematical objects, I become doubtful. My attitude is less a matter of fundamental principles and more a matter of scepticism: I can find no good reason for believing in the existence of the abstract entities postulated by mathematical Platonists. Still, I am willing to be convinced; I have no deeply held conviction that belief in such entities must be avoided at all cost.

The difference in our attitudes can be illustrated by an example from physics. Not long ago, some physicists from Berkeley and Houston made headlines by claiming to have detected a monopole—

supposedly a fundamental particle that some physicists had suspected of existing but no trace of which had ever been discovered prior to that time. This sensational claim was based on a track produced at an enormous altitude in a scientific instrument containing a photographic plate and a system of plastic sheets. The claim was that the track in question could only have been produced by a monopole—no reasonable alternative explanation was feasible. Now it is conceivable that some physicists rejected the claim outright because of a deeply felt conviction that monopoles are mysterious entities the postulation of which must be avoided at all cost. But I suspect that most physicists who were at all sceptical had no such conviction: they simply believed that alternative explanations would be forthcoming. And they were.[15]

The difference in our respective attitudes towards Platonism explains to some extent my willingness (and Gottlieb's unwillingness) to employ modal notions in the exploration of alternatives to mathematical Platonism. I feel that the fact that modal notions may possibly carry hidden commitments to abstract entities should not preclude us from investigating the sort of constructivistic theory developed in this work. For I am concerned with the argument that we *must* postulate abstract entities to have an adequate scientific theory; and my response has always been: "I'm not convinced; perhaps this would do . . .". This sort of response is not vitiated by the claim that perhaps it won't do. Of course, if one regards even the possibility of a hidden reliance on abstract entities as little short of sinful, so that the construction of one's philosophical theories becomes a matter of strenuously avoiding any possible assumption of abstract entities, then I can understand a reluctance to avoid using modal notions. But that has never been my philosophical stance. And in the absence of any convincing reasons for adopting Gottlieb's extreme nominalistic stance—reasons more convincing than just his "intuitions"—I shall no doubt persist in viewing the indispensability thesis as I have all along.

[15] See R. Carrigan, Jr. and W. Trower (Superheavy).

9

Why Burgess Is a Moderate Realist

1. A Dilemma for the Nominalist

JOHN BURGESS has given in (Why) a forceful argument for believing in mathematical objects. Burgess accepts a "moderate version of realism" in mathematics, which is content to observe that our current scientific theories seem to assert the existence of abstract mathematical objects and that we do not yet have good reasons to abandon those theories. He also holds that opponents of his position (all of whom he labels 'nominalists') can take only one of three possible positions in response to this form of realism. These are: (1) instrumentalist nominalism, (2) hermeneutic nominalism, and (3) revolutionary nominalism. But none of these alternatives is, he argues, truly defensible. Thus, he is led to conclude that there is no genuine alternative to accepting the existence of mathematical objects. Of course, this gives only the basic structure of the argument. I turn now to details.

The instrumentalist nominalist adopts an *instrumentalist philosophy of science* "according to which science is just a useful mythology, and no sort of approximation to or idealization of the truth" (Why, p. 93). There is no need for the instrumentalist nominalist to worry about the "ontological commitments" of present-day science, since there is no acceptance of the truth of science. Why worry then about the existential assertions of mathematics? Classical mathematics can be classified as useful mythology too.

Burgess does not regard instrumentalist nominalism as a live option, evidently, for the following two reasons. The first, a slippery slope argument, is that a philosopher who begins by rejecting as outright fiction the mathematically sophisticated theories of physics will find no reasonable place to stop and, in all consistency, should end up rejecting common-sense beliefs as well (Why, p. 94). The second reason is that Burgess thinks that the instrumentalist's professed disbelief in the truth of scientific theories that make reference to mathematical objects is a sham.

This leaves only hermeneutic nominalism and revolutionary nominalism as possible live options for Burgess. Now hermeneutic nominalism is a thesis about common ordinary language. The mathematician and the physicist use ordinary language to make statements about numbers, sets, and functions; they both seem to assert the existence of such things. But, according to the hermeneutic nominalist, when the language is properly analysed and the misunderstandings due to superficial similarities of form cleared up, one will see that the scientist is not actually asserting the existence of any abstract objects.

Burgess dismisses this type of nominalism as a desperate kind of "ostrich nominalism", for he can find no evidence to support the position. As a thesis about the language used by scientists, it should be subject to the evaluation of linguists; but no respectable linguist that he knows of has seriously proposed such a thesis. And no philosopher has adduced any kind of evidence for the thesis that linguists take seriously.

Only revolutionary nominalism remains to be considered. Burgess pictures the revolutionary nominalist as proposing a new kind of science—one in which no existential assertions of the existence of mathematical objects will be found. The problem with this type of nominalism is not that of evidence, as with the previous type, but of reason and motivation. Why ought the scientist to adopt this new type of science? In so far as reasons can be found, they seem to be merely philosophical in nature and hardly sufficient to motivate the kind of scientific revolution being advocated. The only reasonable conclusion to draw is that none of the three alternatives is a live option and that one must accept the fact that science requires us to believe in the existence of mathematical objects. In an (unpublished) earlier version of (Why), Burgess put the idea as follows:

I justify my acceptance of abstract objects thus: Unless and until the physicists tell me that they have revised their theories, I must accept the current ones. Those current physical theories appear to suppose the existence of abstract mathematical objects. Unless and until the linguists tell me that this is an illusion, I must presume that current physical theories really do suppose such objects. Accepting those physical theories, I must accept the abstract entities in question that they posit on pain of inconsistency (Dilemma, p. 22).

He became dissatisfied with this way of expressing his argument, and later put it:

Unless he is content to lapse into a mere instrumentalist or "as if" philosophy of science, the philosopher who wishes to argue for nominalism faces a dilemma: He must search either for evidence for an implausible hypothesis in linguistics, or else for motivation for a costly revolution in physics. Neither horn seems very promising, and that is why I am not a nominalist (Why, p. 101).

To appreciate the force of this argument, let us consider how Burgess treats Field's brand of nominalism. Field is clearly not an instrumentalist nominalist (even though he espouses an instrumentalist philosophy of mathematics), so Burgess concentrates on the two other possibilities. How would Field's nominalism fare when regarded as hermeneutic? Not well at all; it would not have even the slightest plausibility. So Field must be advocating a revolution. But why should scientists make the drastic change being advocated by Field? The only gain would seem to be the avoidance of acceptance of mathematical objects. But is this a gain that scientists would (or should) find significant? Burgess doubts it. He points out, quite rightly, that most working physicists would attach no importance at all to avoiding "ontological commitments" to abstract entities. In any case, he can find no scientific benefits to be obtained from the proposed scientific revolution, and he can see that such a revolution would take place only with considerable costs:

[A]ny major revolution involves transition costs: the rewriting of textbooks, redesign of programs of instruction and so forth. . . . [I]t would involve reworking the physics curriculum, so that each basic theory would initially be presented in qualitative rather than quantitative form. A course on measurement theory would have to be crammed into the already crowded study plan, to explain and justify the use of the usual numerical apparatus (Why, p. 98).

The above criticism is also directed at the predicative view I put forward in (V-C): "A reform along lines of Chihara's . . . would involve reworking the mathematics curriculum for science and engineering students, avoiding impredicative methods in favor of predicative parodies that are harder to learn and not so easy to apply" (p. 98). Thus, Burgess is led to conclude that the nominalist's revolutionary proposal is unscientific and motivated only by "medieval superstition (Ockham's razor)".

Obviously Burgess's argument in favour of accepting a Platonic ontology of abstract mathematical objects is relevant to the views being advanced in this present work. Indeed, it would seem that the specific objections made above to the positions advocated by Field in (Science) and myself in (V-C) can also be applied to my present view. Thus, since I do not accept an instrumentalist philosophy of science, it can be asked if I am advocating a linguistic thesis according to which existence assertions in mathematics are to be analysed as being assertions of the possibility of constructing open-sentence tokens. If I reply in the negative to that question, it would seem that I would have to be advocating a scientific revolution. Then I would be asked to justify such a revolutionary proposal—on scientific grounds. Look at the transition costs. Think of the cost of rewriting textbooks, etc.

2. A False Dilemma

I turn now to an examination of Burgess's reasoning. In effect, Burgess presents his opponents with a dilemma: Are you advancing a linguistic thesis? Or are you advocating a scientific revolution? Either way, your position is implausible. I shall argue, however, that Burgess has presented us with a false dilemma. Notice that no reasons are given for thinking that the alternatives considered above are the only possible ones. Why must Burgess's opponents take one or the other of these horns? Well, what other possibilities are there? How else can one rationally reject Burgess's conclusion? I shall return to this question shortly.

It is worth noting that the kinds of considerations Burgess appeals to in supporting his brand of realism can also be marshalled against the Quinian doctrine (discussed in Chapter 1) that the only kind of mathematical object we have good reason to believe in are sets. For few physicists construct their theories or do their theorizing within the framework of some set theory. How many physicists have learned how to reduce all the various kinds of numbers of classical mathematics to sets? How many physicists have even worked through a development of the fundamental laws of the real numbers in terms of Dedekind cuts? It is not even clear if many physicists regard functions and ordered pairs as sets. For most physicists, it is enough to be able to work with these mathematical objects: they do not have to concern themselves with questions about what these numbers, functions, and ordered pairs really are. Thus, one can argue against the above Quinian position in the way Burgess argued against his anti-realist opponents:

> Why should scientists formulate their theories within the framework of some overarching set theory? The only gain would be ontological: the avoidance of committing oneself to certain kinds of mathematical objects. But is this a gain that scientists should find significant? And look at the transition costs: the rewriting of scientific textbooks, the redesign of programmes of mathematical instruction for science students so that everything would be done set-theoretically. A course on set theory would have to be crammed into the already crowded study plan to explain how the various kinds of objects talked about by classical mathematics can all be reduced to sets.

For a Quinian, the above implications of Burgess's reasoning constitute grounds for rejecting the reasoning—but not for Burgess. He is quite willing to reject Quine's doctrine and commit himself to an ontology of functions, ordered pairs, matrices, Hilbert spaces, etc.[1] However, he does not want to deny that real numbers are Dedekind cuts (or Cauchy sequences or some such mathematical construction). The fact that physicists do not think of real numbers as Dedekind cuts is not crucial. Nor is the fact that the physicist's theories are not formulated in terms of Dedekind cuts. Burgess can allow that there is a kind of "division of labour". It is not necessary for physicists themselves to know what real numbers are. The physicist can work with her intuitive notion of real number, secure in the knowledge that there are mathematicians who have carried out, in a rigorous fashion, the set-theoretical constructions needed to specify objects with those features physicists want the real numbers to have. They can thus rely on mathematicians to tell them what real numbers really are.[2]

But if that is the case, one wonders why Burgess believes that if the anti-Platonist's mathematical constructions are to be theoretically significant, these constructions must enter into such things as the physicist's education, the physicist's textbooks, the physicist's articles, etc. Why should it be necessary to require *physics students* to take courses in type theory in order for the constructibility theory to

---

[1] This was made clear during the discussion of his paper "Epistemology and Nominalism" which was presented at the Philosophy Colloquium of the University of California, Berkeley on 22 Oct. 1987.

[2] This was suggested by Burgess in the discussion mentioned in n. 1, above.

be relevant to the sorts of ontological questions discussed in this book? Why cannot the physicist do for these sorts of questions what she is allowed to do for the case of real numbers? She can simply use the usual mathematical system without worrying about just what one is asserting in making existential assertions in mathematics, secure in the realization that there are foundationalists who can interpret the mathematics they use in a way that does not require that one assert the existence of abstract mathematical objects.

My own position on these matters can be clarified by means of an example. Let us suppose that a society has developed its mathematics within the framework of a Simple Type Theory. We can suppose that the entities treated in the mathematical theory are thought to have all the usual properties that the sets of our standard set-theoretical version of Simple Type Theory are thought to have in our society, except for one small difference: these objects are thought to be astronomical bodies! Hence, what one means (in this society) in asserting the existence of a set of such-and-such a sort is that there is somewhere in physical space an astronomical body with such-and-such properties. Now we can imagine that the scientists in this society do not take the postulation of these astronomical bodies very seriously. For when the mathematicians are asked scientific questions about these bodies, the answers they provide are not very satisfactory from a scientific point of view. How can there be uncountably many astronomical bodies of these sorts in our three-dimensional physical space? "They must overlap each other somehow—think of open intervals that overlap." How is it that we haven't detected any of these astronomical bodies? "They must be in a region of space too far away for us to detect."[3] How have we obtained knowledge of these bodies? "We must have a special faculty—a mathematical intuition—by which we acquire this knowledge." And so on. However, there is no thought of giving up the mathematical theory, because it has proven to be so useful in scientific theorizing. Since all their scientific theories are formulated using the mathematics of this Simple Type Theory, and since the empirical scientists are not trained in logical and

---

[3] To those who believe that these mathematicians have acquired an incoherent theory, I would like to ask: In what way is the belief that mathematical objects are astronomical bodies in a region of space too far from us to detect more incoherent than either the belief of some Platonists that mathematical objects are things that do not exist in physical space or the belief of Penelope Maddy (to be discussed in ch. 10) that sets of physical objects are to be found in the exact location where the physical objects that are its members are to be found?

foundational thinking, there is no attempt to theorize in any other way.

Now imagine that a controversy arises in this society regarding the reasonableness of believing in the astronomical bodies postulated by the mathematicians. Some argue that there isn't any good empirical evidence for the existence of such bodies and that in the absence of evidence, it is not rational to believe in such things. These philosophers come to be known as "ontological Heretics". Others, more orthodox in their beliefs (and hence called "ontological Believers"), respond that there is good evidence for the existence of these bodies: the fact that the postulation of these bodies is essential to our scientific reasoning provides all the evidence needed. Still others, sympathetic to the Heretics, attempt to show by means of complex logical constructions that the postulation of these bodies is not truly required by our scientific reasoning. In this context, a philosopher arises to put forward the Burgess argument for belief in these astronomical bodies:

> Unless he is content to lapse into a mere instrumentalist or "as-if" philosophy of science, the philosopher who wishes to argue for ontological heresy faces a dilemma: He must search either for evidence for an implausible hypothesis in linguistics, or else for motivation for a costly revolution in physics. Neither horn seems very promising, and that is why I am not an ontological Heretic.

But surely there is something very suspicious about such an argument for belief in these astronomical bodies. Would such reasoning provide us with good scientific grounds for belief in such things? Should sophisticated reasoners accept belief in these astronomical objects on the basis of such an argument? I find it hard to imagine very many philosophers or scientists being convinced by the Burgess kind of reasoning. Besides, if one were truly serious in one's belief in these things, then it would seem reasonable to attempt to find out more about them and to develop one's epistemological and metaphysical theories with these objects in mind. How were we able to discover these objects since they cannot be detected by the usual scientific instruments? By what means are we able to determine what properties they have? Why is it that the existence of these things and the relationships which these things have to one another are so important to the empirical sciences? These are just a few of the questions that one would think a serious Believer would pursue. And to believe in the existence of these things just on the basis of the above considerations, without investigating the implications of such a belief for one's epistemological and metaphysical theories, does not strike me as being scientific—on the contrary, it strikes me as being uncritically accepting.

Consider in this light, the constructibility theory developed in this work. It can be used to show how the scientific theorizing of this society could be carried out without assuming the existence of these astronomical bodies. In so using the constructibility theory, one need not be advocating a revolution in the society's physics. After all, since practically none of the scientists in this society takes the belief in the actual existence of the astronomical bodies seriously anyway, there seems to be no harm in continuing to view the mathematical theory as a theory about these strange bodies. And why bother to change textbooks, revise science curricula, etc.? Why not let the mathematicians continue in their present ways, since doing so seems to cause no serious harm? Indeed, it might even be the case that believing in the actual existence of these bodies stimulates these mathematicians to greater devotion to their mathematical work. (It makes their theories seem to them more concrete and important.) Thus, the point of showing these philosophers how mathematics can be done in terms of constructibility quantifiers was not to convert scientists to using a new system of mathematics, but rather to show that the undeniable usefulness of mathematics in science did not require that one believe in the astronomical bodies talked about in mathematics—that there were as yet no strong scientific reasons for believing in these objects.[4]

But does not this response make me into an instrumentalist, which Burgess has already rejected as untenable? Let us recall Burgess's characterization of *instrumentalist philosophy of science*: it holds, according to Burgess, that science does not even provide an approximation to the truth. But this, I surely have not suggested. If, just to sketch one possibility, scientific theories were reformulated within the simple type-theoretical framework I have described in this work, there would be nothing to prevent me from regarding such

---

[4] There was a time when many physicists (e.g. Wilhelm Ostwald) did not believe in molecules, atoms, and the like—they thought that molecules were fictions brought in to aid physicists and chemists in thinking about various forms of energy. What these physicists wanted was scientific evidence for the existence of molecules. But these physicists did not think that we should reform physics so that all mention of molecules should be avoided. For details, see Mary Jo Nye (Molecular, ch. 1).

theories as straightforwardly true. Thus, my view does not preclude allowing that our scientific theories—even those that make use of such heavyweight mathematical theories—are at least close approximations to true theories. (I shall say more about this point in Chapter 12.)

The main difference that separates us then comes to this: I want to treat the question of the existence of mathematical entities as a scientific question, whereas Burgess wishes simply to accept as literally true in all respects current scientific theories and to dare his opponents to justify reforming science. Thus, I don't at all agree with him that I am proposing some new metascientific viewpoint from which to judge the veracity of scientific theories. On the contrary, I wish to use the usual scientific standards to assess the plausibility of the realist's claims. It is here that I find the realist's claims so unsatisfactory. Where is the evidence that supposedly supports belief in these entities? Here, I mean evidence of the sort that would convince the physicist or biologist. Certainly, Burgess supplies none at all. So it is hard to see why he is so confident that these mathematical objects exist. And even if he is able to provide some kind of philosophical reason for holding on to such a belief, despite the absence of any scientific evidence, it is hard to see why he should claim that his position is the scientific one and that his opponent's position is superstitious and dogmatic.

It should be clear then how I would respond to Burgess's dilemma for the nominalist. I accept neither of the alternatives he allows his opponents. What I do advocate is the assessing of claims made by philosophers that we ought to believe in the existence of entities that do not exist in the physical world. (Notice that such claims are not made by scientists: in fact, most scientists with whom I have discussed such questions find the Platonist's claims about the existence of sets to be either unintelligible or meaningless.) And I do advocate using the same sorts of standards of assessment and evaluation that scientists use in evaluating the plausibility and cogency of similar sorts of claims regarding the existence of physical entities that cannot be observed in any direct way. It is in connection with such evaluations that I produce alternative theories of mathematics that do not, at least in any obvious way, assert the existence of mathematical entities. If science can be done with these alternative theories, what evidence do we have that the existence of Platonic entities is required by the scientific evidence we have acquired? And if there is no good positive evidence for their existence, then it is reasonable to be somewhat sceptical about the ontological theories spun out by the TRUE BELIEVERS in mathematical objects.

As for Burgess's brand of realism, it seems to me that it is hardly moderate. For Burgess seems to commit himself to including in his ontology not only a tremendous variety of mathematical objects (such as numbers, sets, matrices, functions, operations, vectors, fields, groups, rings, etc.) but also such *things* as chances, difficulties, possibilities, solutions, and differences. Thus, when I read works in physics (both textbooks and research papers), I find such sentences as 'There are different possibilities for describing the space M . . .', 'There are as many problems connected with that hypothesis as there are with this one', and 'There are a great many serious difficulties connected with this theory.' So imagine someone arguing as follows:

> I justify my acceptance of objects called possibilities, problems, and serious difficulties thus: unless and until the physicists tell me that they have revised their theories, I must accept the current ones. Those current physical theories appear to suppose the existence of such abstract objects as possibilities, problems, and serious difficulties. Unless and until the linguists tell me that this is an illusion, I must presume that current physical theories really do suppose such objects. Accepting those physical theories, I must accept the abstract entities in question that they posit on pain of inconsistency.

It might be said that the existential statements about difficulties, chances, solutions, and differences made by physicists are not serious expressions of their theoretical beliefs, but instead are mere colloquial expressions of thoughts which should not be taken seriously. Thus, there is no good reason for us to believe in such strange entities as difficulties and chances, as there is in the case of numbers and functions.[5]

But how are we to determine what is, and what is not, a serious statement of the physicist's theories? If a physicist states that Avogadro's number is greater than 6×10^{23}, we are to take the scientist to be expressing a theoretical belief about abstract mathematical objects which we ought to take seriously. On the other hand, if a

---

[5] Burgess made such a reply during the discussion of his paper "Epistemology and Nominalism" mentioned in n. 1, above.

physicist states that there is a distinct possibility that Avogadro's number is greater than 6×10^{23}, we are to treat the statement as loose talk that should not be taken seriously. But by what criteria are we to determine what is, and what is not, a serious expression of the physicist's theoretical beliefs? After all, if we asked the physicist if she really meant what she said when she said 'There is a distinct possibility that Avogadro's number is greater than 6×10^{23}', she would probably reply that she did. She might even reply that she was more certain of that claim than she is of the statement that Avogadro's number is greater than 6×10^{23}.

Imagine that a meteorologist issues a report in which it is stated that there is a strong chance of rain in the near future. It would be strange to maintain that the meteorologist's statement is not the expression of one of her theoretical beliefs. It would be even stranger to regard this statement as loose talk that we should not take seriously.

Burgess once responded (in a letter) to the above point by suggesting that the physicist should be asked to paraphrase such statements to see if she really is presupposing such entities. It needs to be kept in mind that Burgess should not be saddled with a view he has expressed only in a personal letter. However, I should like to discuss this response because doing so will, I believe, prove to be illuminating. First of all, it needs to be noted that the physicists who wrote such things may not be in a position to paraphrase these statements: they may no longer be alive. Secondly, why should we ask these physicists? After all, in criticizing the nominalist, Burgess demanded that we go to the linguist. He demanded linguistic evidence for paraphrasing the statements of science. Should he not do the same here? And should he not demand that the physicists doing the paraphrasing provide evidence that the paraphrases are correct (as he did of the nominalist)? Incidentally, is there any reason to think that physicists have any expertise in doing this sort of thing? Take the second of the sentences I gave above as examples. How many physicists can be expected to give a decent paraphrase of this which would not seem to presuppose the existence of problems? In short, if we understand the paraphraser to be proposing a linguistic thesis (hermeneutic nominalism), the paraphraser will be faced with the sorts of difficulties that Burgess outlines in his paper.

Of course, Burgess can take the physicist to be a kind of revolutionary nominalist, advocating a reform of physics textbooks and the like. Then shouldn't these physicists be asked to show how there will be a scientific gain in reforming science by paraphrasing such sentences in the way being advocated? To put it another way, why don't the physicists face the dilemma they put to the nominalist?

3. Burgess's Objection to Field's Preference for Nominalism

I shall end with a few words about the two short objections Burgess raises to Field's suggestion in (Science), p. 98, that there are epistemological reasons for preferring nominalism to realism. Field had argued:

[Nominalism] saves us from having to believe in a large realm of . . . entities which are very unlike the other entities we believe in (due for instance to their causal isolation from us and from everything that we experience) and which give rise to substantial philosophical perplexities because of those differences (Science, p. 98).

Burgess's first point of rebuttal is that "as Maddy has skilfully argued in her (Perception), there is a good deal of research in developmental psychology and neurophysiology that can be read as showing that we do, in a sense, have causal contact with certain abstracta" (Why, p. 100). I shall say nothing here about this first point, because I shall provide a detailed evaluation of Maddy's reasoning in the next chapter. Burgess, however, has a second objection to Field's reasoning:

Suppose that Burrhus Skinner were to confess that after all those years of work with his rats and pigeons he's still "substantially perplexed" by the ability of freshman students to master calculus and mechanics. . . . No one would take it as an indication of anything but the inadequacies of behaviorist learning theory.

Likewise, a philosopher's confession that knowledge in pure and applied mathematics perplexes him constitutes no sort of argument for nominalism, but merely an indication that the philosopher's approach to cognition is, like Skinner's, inadequate (Why, p. 101).

Now is this an appropriate response to Field's reasoning? As I see it, Field's principal point can be put in the form of a conditional: *If* it is supposed that knowledge in mathematics consists in the knowledge of the existence and properties of certain objects which do not exist in the physical world and with which we have no causal interaction at all, *then* mathematical knowledge would be a complete mystery. After all, the inference that Field wants us to draw is not that mathematical knowledge is inexplicable, but rather that the above supposition about mathematical knowledge is false. (An exposition of Field's "deflationist" view of mathematical knowledge will be given in Chapter 12.) The Skinner analogy is thus misleading. It would be more accurate to compare the situation Field is describing with the following: Imagine that it is held by some linguists that whether any grammatical rule of English is correct or not depends on what is written on some stone tablets and that these stone tablets are to be found in another universe with which we have no causal interaction. It is then argued by some philosopher that if these linguists were right, then it would be a complete mystery how we could ever know that any grammatical rule of English we formulated was in fact correct. Would it be reasonable for the stone-tablet linguists to reply to this philosopher in the way Burgess replies to Field? "This philosopher's confession that our knowledge of grammatical rules of English perplexes him constitutes no sort of argument for rejection of our position, but merely an indication that the philosopher's approach to cognition is inadequate."

10

Maddy's Solution to the Problem of Reference

1. Involved Platonists

PENELOPE MADDY is an Involved Platonist. What I mean by this can best be seen by reconsidering the example, described in the previous chapter, of the society in which everyone is taught that the existential theorems of their mathematics assert the existence of astronomical bodies. Recall that in this society there are ontological Believers and ontological Heretics, that is, those who believe in the existence of these astronomical bodies and those who do not. Among the ontological Believers, I would like to distinguish *Unthinking Believers, Detached Believers*, and *Involved Believers*. The Unthinking Believer believes in the existence of the astronomical bodies postulated by the mathematicians because that is what she has been taught; she does not question this belief or consider the consequences of this belief. The Detached Believer, on the other hand, comes to her belief in the existence of these bodies as a result of reflection and reasoning. In particular, she accepts belief in these postulated astronomical bodies because the opposing heretical position seems to her to be too difficult to maintain; but she herself does not develop theories about the nature of the bodies she believes in. She stands detached from the theoretical problems that her belief in the astronomical bodies engenders. Finally, the Involved Believer treats these postulated bodies as she would the postulated entities of physics and chemistry. She theorizes about the nature of these astronomical bodies and attempts to draw conclusions about both the properties of these postulated bodies and also the nature of our minds. Epistemological difficulties connected with the postulation of these bodies are not mere philosophical playthings to the Involved Believer, to be ignored or scorned, but instead are treated as serious theoretical problems to be resolved by reason and theory construction.

Michael Resnik and Penelope Maddy are like the Involved Believers of the example. These two Platonists attempt, by theory construction and reason, to deal directly with the deep epistemological and theoretical difficulties that their belief in mathematical objects produces. Earlier, we saw how Resnik attempted to devise an ontological and epistemological theory that would obviate some of the serious philosophical problems of mathematical Platonism. In Section 3, I shall sketch the ontological theory that Maddy has developed in response to a problem for mathematical Platonism that has recently come to light.

2. Problems of Reference for Mathematical Platonism

As an introduction to this problem, I should like to develop further the above example of the society whose mathematics is taken to be a theory about certain astronomical bodies. Imagine that some philosophers of this society begin to wonder just how it is that mathematicians were able to pick out any particular astronomical body as the referent of the term 'the empty set'. After all, these mathematicians do not claim to know anything about the physical properties of the bodies to which mathematical terms refer: they do not know where they are to be found, what sizes they have, or what they look like. No one claims to have actually pointed at one of these astronomical bodies and named it 'the empty set'. No one knows how to pick out, from among the totality of astronomical bodies, that particular one that is supposed to be the empty set. Given this fact, how can one still maintain that the term 'the empty set' refers to some particular astronomical body?

A problem of this sort for mathematical Platonism in our world has been posed by Jonathan Lear in his (S and S). Lear suggests that we are all Platonists "at first blush": we all believe that statements of set theory are *about* sets. We also believe, he claims, that sets are abstract objects of some sort, which do not exist in space and time. This naïve Platonic view can then be characterized by the following doctrines: (1) sets are abstract objects; and (2) set-theoretic discourse is about these abstract objects (S and S, p. 88). Then he poses the problem: How is it possible that we succeed in talking about these abstract objects? This is a problem, it would seem, because we cannot begin talking about sets in the way we can begin talking about, say, gold. In the case of gold, we can stand in a causal relation with samples of gold: there can be initial acts of "dubbing"—"Let's call stuff of this kind 'gold'" (pointing to some samples of gold). According to some recent theories of reference derived from the writings of Saul Kripke and Hilary Putnam,[1] we succeed in referring to and talking about the kind *gold* in virtue of standing in a causal chain that began with such a dubbing. Since it appears that we do not stand in any sort of causal contact with the abstract objects called 'sets', Lear concludes that the Platonist must develop an account of reference according to which one can succeed in talking about a kind of thing despite the absence of any causal contact with things of that kind (p. 102).

I would like to develop Lear's problem, but in a way which is somewhat different from what is found in his paper. Suppose, for the sake of argument, that there is a totality, T, of abstract objects which are related in the way we think the pure sets (of the iterative conception of sets) are related.[2] In other words, there is a relation R over T which has the structural features we attribute to the membership relation. It follows that there is an object in T to which none of the objects in T is in the R relation. This object has the structural properties we attribute to the empty set. Call this object 'zero'. Now consider the totality of all those objects in T which is formed in the following way. We first define a new relation R* in terms of the relation R as follows: For all objects a and b,

$aR*b$ iff a is different from zero and aRb.

Since there is an object in T—call it 'zero*'—to which only zero is related by the R-relation, no object in T is in the relation $R*$ to zero*. We now build up a structure from zero* in the way the standard structure is built up from zero according to the iterative conception. We do this in stages. At stage 0, we have only zero*. At stage 1, we have the object to which only two objects are related by R, namely zero and zero*. In other words, we have that object to which only zero* is the $R*$relation. Call this object: {zero*}*. At stage 2, we will have: zero*, {zero*}*, {{zero*}*}*, and {zero*, {zero*}*}*. And so on. In general, at stage $x + 1$, we get all the objects that correspond to *subsets of the totality of objects formed at stages earlier than x*, according to the standard iterative picture. But these "subsets" would

---

[1] See especially Kripke (Naming) and Putnam (Meaning).
[2] See George Boolos (Iterative) for an extended discussion of this conception.

be formed by taking what would be a subset, if R were the genuine membership relation, and adding zero as a "member". At limit ordinals, one gets all the objects formed at earlier stages. One continues this process as one does according to the iterative conception. The totality one gets in this way will be called T^*.

Clearly, $< T^*, R^*>$ is structurally indistinguishable from $\langle T, R \rangle$. And within the framework of the structure $\langle T^*, R^* \rangle$, zero* will have all the structural properties that zero has relative to the structure $\langle T, R \rangle$. It would be a simple matter to show, by a similar process, that there must be infinitely many such structures that are isomorphic copies of $\langle T, R \rangle$. How then do we succeed in picking out some *particular* totality of abstract objects as the universe of sets? How do we succeed in picking out some particular abstract object as the referent of 'the empty set'? Merely describing structural features of this totality clearly will not do.

It may be thought that one could pick out $\langle T, R \rangle$ from the multitude of isomorphic copies described above, since one could determine which structure is the one from which all the others were generated. Could one not see, for example, that every member of T^* is a member of T, but not that every member of T is a member of T^*? But there are many problems with this suggestion. First of all, why can we assume that each of the "copies" described above would be presented to us in a form that would relate it to $\langle T, R \rangle$? After all, we cannot literally see the members of T and T^*. Besides, even if one could somehow pick $\langle T, R \rangle$ out of this crowd, how would we know that T is the totality of sets described by the iterative conception? The assumption with which we began was only that T is a totality of objects related in the way we think the pure sets of the iterative conception are related by the membership relation. This assumption does not guarantee that R is the genuine membership relation.

Mathematical Structuralists will not see the above as a problem, since structure is all that is important to them. But such considerations as the above do pose a pressing problem for Maddy, because she regards sets as unique objects that form a natural kind: "[F]rom the point of view of the set theoretic realist, the treatment of sets as forming a kind is much more likely to be correct than a more traditional theory according to which a set is anything which satisfies certain conditions" (Perception, p. 183). Thus, Maddy is led to wrestle with the question of how we are able to refer to objects of this kind. How can we refer to things that we cannot see, touch, detect, or locate in physical space? How can we refer to things from which we are so totally isolated?

3. Maddy's Solution

Maddy attempts to obviate problems of the above sort by claiming that we do stand in causal contact with sets. The principal idea underlying her resolution of this problem of reference to sets is that sets (at least certain sets) are very much like physical objects— according to Maddy's theory of sets, certain kinds of sets, namely sets of physical objects (such as the set of eggs in my refrigerator), have all the usual properties of physical objects: they have location in space and time, they can be destroyed, brought into existence, seen, touched, moved about, tasted, etc. Thus, the mystery of how we can refer to these objects is solved. The case is no different, it would seem, from the case of gold. We can dub sample sets: "This is the unit set whose only element is this apple". Maddy concludes that the kind of entity to which we refer when we use the terminology of set theory, that is, the kind *set*, can be dubbed by picking out samples; and she then asserts: "Particular sets and less inclusive kinds can then be picked out by description, for example, 'the set with no elements' for the empty set, or 'those sets whose transitive closures contain no physical objects' for the kind pure sets" (Perception, p. 184). Notice that Maddy is, in effect, putting forward the view that we refer to particular abstract objects when we use such terms from set theory as 'the set with no elements'. Evidently, she believes that within the totality of all abstract objects, there is some particular object that is the set with no elements and that it is this particular object to which we refer when we use the expression 'the set with no elements'. How, according to Maddy, are we able to achieve these feats of reference? Roughly, it is supposed to work this way: We are supposed to be acquainted with certain sets; these we can dub 'sets' so that we can use the word 'set' to refer to the kind *set*. Then we can use definite descriptions to pick out, from the totality of objects of this kind, particular sets to refer to.

How does Maddy arrive at her doctrine that sets of physical objects have location in space and time? In this regard, we do not get argumentation but rather mere assertion. Thus she writes

I must agree that many sets, the empty set or the set of real numbers, for example, cannot be said to have location, but I disagree in the case of sets of physical objects. It seems perfectly reasonable to suppose that such sets have location in time—for example, that the singleton containing a given object comes into and goes out of existence with that object. In the same way, a set of physical objects has spatial location insofar as its elements do. The set of eggs, then is located in the egg carton—that is, exactly where the physical aggregate made up of the eggs is located (Perception, p. 179).

Notice that Maddy distinguishes the set of eggs from the "physical aggregate" made up of the eggs. The latter seems to be what is sometimes called the *mereological sum* of the eggs.[3] Hence, each egg is related to the physical aggregate in a way that is quite different from the way it is related to the set of eggs: each egg is made up of physical material that forms the parts of the physical aggregate; but none of the eggs is a member of the physical aggregate, for a physical aggregate does not have members. It is the set of these eggs to which each egg is related by the membership relationship. And this set, according to Maddy, is "an abstract object";[4] the physical aggregate is not. However, the set of eggs is supposed to be located in exactly the place the physical aggregate is located.

It needs to be emphasized that, for Maddy, the set of eggs is a distinct entity from the physical aggregate whose parts make up the eggs, as can be seen from the following quotation:

A more difficult question is why the numerical perceptual beliefs in question should not be considered to be beliefs about the physical aggregate, not the set. These beliefs are beliefs that something or other has a number property, and Frege has soundly defeated the view that a physical aggregate alone can have such a property (Perception, p. 179).

I infer from this passage that Maddy believes that a cardinality statement, such as 'There are three apples on the table', implies that something or other has a number property. Evidently, she also believes that no physical aggregate can have such a property (since "Frege has soundly defeated . . ."), whereas sets can. It follows, according to Maddy, that no physical aggregate can be a set.

And what justification does Maddy provide for her claim that we can perceive these sets of physical objects? First, she sets forth an

---

[3] Above, ch. 6 for more on mereological sums.
[4] "I have now denied that abstract objects cannot exist in space and time, and suggested that sets of physical objects do so exist" (Perception, p. 179, n. 39).

account of perception, derived from George Pitcher, according to which:

Person *P* perceives an object of kind *k* at location *l* if, and only if,
  (i) there is an object of kind *k* at *l*;
  (ii) *P* acquires perceptual beliefs about the kind *k*, in particular, that there is an object of kind *k* at *l*;
  (iii) the object at *l* is involved in the generation of the perceptual belief state in an appropriate causal way.

She then asserts (with very little argumentation or empirical support) that it is reasonable to assume these conditions are met in cases in which the objects of perception are sets of physical objects, postulating in the adult human brain a neural 'set detector':

[W]hen *P* looks into the egg carton, (i) there is a set of eggs in the carton, (ii) *P* acquires some perceptual beliefs about this set of eggs, and (iii) the set of eggs in the carton is appropriately causally responsible for *P*'s perceptual belief state. The involvement of the set of eggs in the generation of *P*'s belief state is the same as that of my hand in the generation of my belief that there is a hand before me when I look at it in good light . . . As in the case of knowledge of physical objects, it is the presence of the appropriate detector which legitimizes the gap between what is causally interacted with, and what is known about (Perception, pp. 182–3).

It is obvious that Maddy's ideas about sets are strikingly different from those of Resnik's. Although both of these Involved Platonists believe in mathematical objects, the properties that they attribute to these objects are radically different. Mixed sets, for Maddy, are not mere positions in a structure: they have properties, such as position in physical space, which are independent of their set-theoretical properties. Furthermore, it does make sense, according to Maddy's theory, to ask if some set from one structure is or is not identical with some set from another structure. On the other hand, I have learned from Resnik that he does not believe (as does Maddy) one can see sets or move sets about.

It may also be helpful at this point to contrast Maddy's views about sets with those of Gödel's. Maddy holds that certain kinds of sets have location in physical space and time and can be literally perceived by us. Gödel did not maintain that we can literally perceive sets, but only that we have something like a perception ("mathematical intuition") of the objects of transfinite set theory—objects which he believed do not exist in the physical world.

4. Doubts about Maddy's Solution

Let us consider some of the consequences of Maddy's views. Suppose that I have completely cleared the surface of my desk, leaving nothing but a single apple. According to Maddy, we may think that there is nothing else on the desk, but in fact, there is something else there, namely the set whose only element is the apple. Remember, the physical aggregate made up of the apple is one thing; the set whose only member is the apple is another thing. The singleton is an abstract object; the physical aggregate is not. But the singleton is located in exactly the same place as the apple. Indeed, according to Maddy, the set came into existence at the exact moment in time that the apple came into existence, and it will go out of existence at the exact moment that the apple goes out of existence. And this object can be moved from place to place. How? When one moves the apple, the unit set gets moved automatically. Now what does this set look like? Evidently, it looks exactly like the apple. After all, I cannot see anything else there on my desk that looks different from the apple. Perhaps, then, this set feels different. But when I put my hand where the set is supposed to be, what I feel is no different from what I feel when I put my hand on the apple. Well, does the set taste different from the apple? Take a bite and see. No, the set tastes just like an apple. We thus have an answer to the riddle: What looks like a duck, waddles like a duck, quacks like a duck, smells like a duck, tastes like a duck, . . . but isn't a duck? It's the set whose only member is the duck.

Actually, the situation is worse than it may appear at this point, for it is not just the unit set whose only element is the apple that is on my desk. Since this unit set has location in space, can be perceived, etc., we can infer that the unit set whose only element is the unit set also has location in space, can be perceived, etc. For Maddy accepts the following principle: If objects A, B, . . . have location in space, then the set of these objects has the same location in space.[5] It follows that there is on my desk, not only the apple and the unit set whose only

---

[5] Maddy did not explicitly espouse such a principle in (Perception), although she strongly suggested that she was presupposing some such principle, writing: "[A] *set of physical objects has spatial location insofar as its elements do*" (Perception, p. 179, italics mine). However, in a more recent (unpublished) manuscript, she explicitly advocates such a principle, stating: "[A] set whose transitive closure contains physical objects is located where the aggregate of physical stuff making up its members (and the members of its members, etc.) is located."

element is the apple, but also the unit set whose only element is this unit set whose only element is the apple. Clearly, by this line of reasoning, we can infer that there are infinitely many such objects on my desk. And all these different objects take up exactly the same amount of space on the desk. Furthermore, this infinity of abstract objects came into existence when the apple came into existence, and they will all go out of existence when the apple goes out of existence.

Lest it be supposed that all these strange consequences flow from the peculiarities of unit sets, imagine that there is a cup and saucer on my dining-room table, and nothing else there. By the above principle, there is in the exact location of the cup and saucer in addition to the physical aggregate an infinity of other distinct objects: there is the unordered pair consisting of the cup and the saucer; there is the unordered pair consisting of the cup and the above unordered pair; there is the unordered pair consisting of the saucer and the above unordered pair; etc. And all of these distinct objects occupy exactly the same space.

These considerations give rise to the following inquiry. Most people I know do not claim (literally) to see sets. Maddy is the only philosopher I know to have made such a claim. Yet many believe in the existence of sets. Many mathematicians, for example, believe that the eggs in the carton are members of the set of eggs in the carton. Such a person could see the eggs as members of the set of eggs in the carton, even though she does not believe that the set is something one could see. Presumably, such a person could count the eggs and, as a result of this counting, could come to believe the very same cardinality statement as Maddy would come to believe as a result of such a counting, namely that there are twelve eggs in the carton. So the question arises: What would such a person be lacking? That is, the person who sees the eggs as members of the set of eggs in the carton, but who does not see the set, differs from the person who both sees the eggs as members of the set of eggs in the carton and also sees the set. What does the latter have the former does not? Does the former lack the neural set detector that Maddy postulates? Or is the set detector just not functioning in the case of the former? What would the former have to do to be able to see the set? Does she need training in set detection? Imagine a course in set detection. What would such a course be like?

In the above, I have assumed that Maddy would maintain that the unit set whose only element is the apple on my desk *looks exactly like the apple*. This seemed to me the only reasonable position for Maddy to take, given her basic hypothesis that this set occupies the exact location of the apple. For how can something that occupies the exact location of the apple and that reflects light in exactly the way the apple does not look like the apple? Maddy, however, has informed me that, for her, the unit set in question does not look exactly like the apple.[6] When she perceives the set, she claims, her perceptual experience is different from the one she has when she perceives the apple.[7]

Imagine that physicists, armed with the most sophisticated of instruments, attempt to detect the unit set in question. What would their instruments tell them? Would they pick up any signs of the set? For example, would they detect light rays emanating from the set that cannot be attributed to those emanating from the apple? Of course not. The set cannot be detected in this way. How is it then that the set looks different from the apple?

Maddy's claim that the unit set looks different from the apple is based on the fact that the perceptual experience she has when she sees the set is different from the one she has when she sees the apple. But is this difference in perceptual experience sufficient to justify her claim? Imagine that a piece of ply wood is cut into the shape of Kohler's famous duck-rabbit (discussed by Wittgenstein in part II of his (Investigations)). Suppose that I show this piece of wood to my daughter. She exclaims: "Oh, what a cute rabbit!" I say: "It's a duck— look, here is its bill." When she sees the piece of wood as a duck, her perceptual experience is different from the one she had when she saw it as a rabbit. Since it is possible to have one sort of perceptual experience when one perceives the piece of wood as a rabbit-shaped object and to have a different sort of perceptual experience when one perceives it as a duck-shaped object, it follows that such a difference of perceptual experience does not justify the kind of conclusion Maddy wants us to draw, namely that the object my daughter perceived when she saw the duck-shaped thing *must look different* from the object she perceived when she saw the rabbit-shaped thing. For the object my daughter perceived when she saw the duck-shaped thing is identical to the object she perceived when she saw the rabbit-shaped thing; and hence the object my daughter perceived when she saw the duck-shaped thing looks exactly like the object she perceived when she saw

[6] This was pointed out to me in a letter dated 8 Sept. 1987.

[7] This view is only suggested in (Perception), but she explicitly espouses such a doctrine in her (Reply).

the rabbit-shaped thing. I am assuming here that if A = B, then A looks exactly like B. Or, as David Kaplan once put it, "whatever you may think about the identity of indiscernibles, *no* sensible person would deny the indiscernibility of identicals" (Heir Lines, p. 99).

To reinforce these reasons for questioning the principle upon which Maddy is relying, let us imagine that I am sitting on my front deck when I see my friendly next-door neighbour speaking to my daughter across the fence. I have always considered my neighbour to be a kindly old gentleman. However, as they are talking, my wife hurries up to me and informs me that she has just learned from the local police station that our neighbour is in fact a registered sex offender, who has sexually molested many children. The perceptual experience I now have as I look at my neighbour talking to my daughter is significantly different from the experience I had a moment earlier. But can we infer from this difference in perceptual experience that my neighbour has changed his appearance? Surely, it is only my own (subjective) reaction to seeing him talking to my daughter that has changed. I now see him differently. Maddy's claim that she can see the unit set whose only element is the apple and that this unit set looks different from the apple is based, I conclude, on very questionable reasoning.

Maddy claims that sets of physical objects can be seen, touched, tasted, and so on. But even supposing that sets of physical objects occupy the exact location in space that the physical objects do, it is not at all obvious that we should say such sets are visible. Thus, imagine that a philosopher proposes a theory according to which each person has a *doppelgänger*—another person, who lives, breathes, and acts just like the person in question. Unfortunately for this philosopher, no empirical evidence is ever uncovered to indicate the existence of any *doppelgängers*. Then one day, this philosopher hits upon a theory that she thinks will guarantee that there will be direct visual evidence for the existence of these *doppelgängers*. She proposes that each person's *doppelgänger* occupies the exact location in space that the person does. In that case, she reasons, we can gather empirical evidence for the existence of *doppelgängers*: These *doppelgängers* will be visible. And, in response to the objection that these *doppelgängers* will be indistinguishable from the "originals", she replies: "Oh no—a person's *doppelgänger* does not look exactly like the person: the perceptual experience I have when I see a *doppelgänger* is different from the one I have when I see the person."

Of course, no scientist would take such a theory seriously. Most would continue to maintain that there is no empirical evidence for the existence of *doppelgängers*. They would insist that their instruments had not detected any signs of the *doppelgängers* and they would discount claims to have actually seen these things.

The above considerations point to another possibility. Suppose that there is a society of mathematicians who believe, in contrast to Maddy, that what occupies the exact location of my apple is not the unit set, but rather the pair set whose only elements are the apple and the null set. Furthermore, where Maddy would say that $\{a, b, c, \ldots\}$ occupies the exact location of the physical aggregate consisting of a, b, $c, \ldots$, these mathematicians claim that $\{a, b, c, \ldots, \text{the null set}\}$ occupies this location. In other words, whereas Maddy believes that sets of physical objects exist where its members are, these mathematicians believe that it is sets* of physical objects that so exist. We can imagine that these mathematicians believe that a set which does not contain the null set as a member is like a body without a soul: "Such sets just could not exist in space," they say.

Needless to say, physicists would not be able to determine by means of all their instruments which of these competing hypotheses is right. Notice that the relation that obtains between all, and only, the apples on my desk and the set of apples on my desk is the membership relation; whereas the relation that obtains between all, and only, the apples on my desk and the set* of apples on my desk is the membership* relation, where x is a member* of y if, and only if, x is different from the null set and x is a member of y. Obviously, nothing of any practical consequence would be changed if one adopted the society's hypothesis instead of Maddy's. Whenever Maddy would say that some egg is a member of the set of eggs in her refrigerator, these mathematicians would say that this egg is a member* of the set* of eggs in the refrigerator. All of our ordinary cardinality judgements could be analysed in terms of either hypothesis with no significant differences. Indeed, I believe that most scientists would find the two hypotheses structurally identical and equally plausible (or implausible).

But there would be philosophical consequences. If the mathematicians we are considering were right, then Maddy would be making many mistakes. For example, she would be taking what was in fact the membership* relation for what she believed was the membership relation. And what was thought by her to be dubbed 'the set of apples on my desk' would in fact be the set* of apples on my desk. Thus, the dubbing ceremony would not have succeeded in selecting the referential target that Maddy was after.

Perhaps it will be said in reply that *it just could not be the case that what exists in space are sets* of physical objects*. But why? Take Maddy's hypothesis that what exists in space are sets of physical objects but not sets*. Why should Maddy's hypothesis be possible but not the mathematicians'? I can find no a priori demonstration of any such modal thesis (and certainly Maddy has provided no such demonstration). What about an a posteriori justification of the thesis? Surely no one has conducted an empirical investigation that establishes her modal thesis.

Thus, until Maddy provides us with convincing evidence for her claim that what exists in space are sets of physical objects and not sets* of physical objects, she cannot claim to have truly resolved the problem of reference with which we began this chapter. She has not satisfactorily explained, *even on the hypothesis that there are sets*, how it is that we have succeeded in referring to the empty set (as opposed to, say, the empty set*).[8]

In the above discussion, I have considered only one rival hypothesis to the one Maddy has put forward. There are clearly many others we might consider. What if what exists in space are sets** of physical objects? Or sets*** of physical objects? (I leave it to the reader to supply the appropriate definition of 'sets**', 'sets***', and so on.) There are countless possibilities. But there is one other possibility (which I shall call 'the null hypothesis') that needs special attention. *Suppose that no sets at all exist in physical space*. One can still maintain that it is sets (or sets* or sets** . . .) that have number properties. One can still apply cardinality theory in the way Logicists have outlined: One can still determine by counting that the set of pictures in my living-room has cardinality ten; and there is no obvious reason why such a judgement requires that one literally see the set itself or that the set exist in the room itself. And the null hypothesis does not conflict, in any obvious way, with the common perceptual experiences we have all had of seeing that there are three books on some table: there is no obvious reason why we could not see

---

[8] Cf. Maddy's description of her own goals in (Perception): "I will not attempt to prove, or even argue for, the independent existence of sets, but rather, on the realistic assumption that they do so exist, I will try to show how we can refer to and know about them" (p. 165).

that there are three books on the table without seeing the set itself. Thus, consider the example Maddy gives in (Reply) of a biologist seeing three paramecia and an amoeba, where an artist sees only a microscopic image that looks like an undifferentiated mass of Jackson Pollock drips. Maddy analyses this case to be one in which there is a real difference in the phenomenological look of a given scene and claims that there is a difference in perceived objects: the biologist sees a set, whereas the artist sees a physical mass. But why must the biologist have seen a set? Even if we grant for the sake of argument that a cardinality judgement of the above sort is a judgement *about* a particular set, why must the set be seen in order that the judgement be made? The biologist sees the paramecia and the amoeba, and she sees that there are three of the paramecia and one of the amoeba. Why must she have seen the set?

It may be thought that one must see a set in order to see (directly) that it has three elements, just as one must see a jar in order to see (directly) that there are three cherries in it. How could you tell just by looking that the jar contains three cherries if you could not see the jar? Similarly, it might be thought that one needs to see a set in order to know what its cardinality is. Behind this reasoning is an idea many people acquire that the members of a set are "in the set" in the way the cherries are in the jar. But of course a little reflection will reveal that such an idea is unwarranted. To say that a cat is in the set of cats is just another way of saying that the cat is a member of the set of cats. We may use the familiar braces to construct names of sets, and we frequently write the names of various objects *inside* braces to indicate that the objects are members of the set specified by the braces. Still, even though the names of the objects occur inside the braces, we cannot infer that the objects named are to be found literally (and spatially) inside the set.

To say a cat is in the set of cats is just another way of saying that the cat is a member of the set of cats. Well, what is the relationship that obtains between each cat and the set of cats, in virtue of which each cat is a member of the set of cats? Recall that the crucial difference between the set of cats and the physical aggregate made up of all cats is that each and every cat is related by the membership relation to the set of cats, but no cat is related by the membership relation to the physical aggregate. The set of cats is that set to which a thing is in the membership relationship if, and only if, the thing is a cat. We know what things are related to this set by the membership relationship as a result of being told that this set is that set to which each and every cat (and nothing else) is related by the membership relation. More generally, when we speak of the set of *F*s, we mean that set to which a thing is related by the membership relation if, and only if, the thing is an *F*. Notice that it is the condition *F* that does all the work: we don't have any independent way of determining if a thing is a member of the set.[9]

Thus, suppose I point at a jar and say: "What pieces of fruit are inside the jar?" You would not have to be told what pieces of fruit are related to the jar by the *is inside of* relation in order to know what jar is being referred to (imagine that the jar is made of an opaque glass); you could know what jar is being pointed at and then determine for yourself what pieces of fruit are inside the jar. But the case of sets is different. In order to know what set is being referred to, you need to be given somehow the things that are in the membership relation to it. And once you are given that, *that* will tell you what things are in the membership relation to it.

Thus, in order to tell, just by looking, that the set of books on my desk has cardinality eight, I need to see the books. And I may need to count the books. But why do I need to see the set? Do I need to see the set in order to know that the books I see on my desk are members of the set? Surely not. For I know that something is a member of this set if, and only if, the thing is a book on my desk. Then what reason do we have for insisting that I must see the set? What reason does Maddy have for insisting that the biologist *sees the set* of paramecia and amoeba on the slide? It is clear, from the analysis of cardinality given in Chapter 5, that seeing there are eight books on the desk does not logically presuppose that the set of books on the desk be seen or even that this set exists.

It is now evident that Maddy has relied upon one of an infinity of possible hypotheses, without anything remotely like convincing grounds for picking her favoured hypothesis out as the right one. Lacking these grounds, we see that there are reasons for seriously questioning Maddy's solution to the problem of reference: in particular, she has not given a satisfactory answer to how it is that we have succeeded in picking out a unique abstract object and referring to it as the empty set. Thus, even if sets exist, how are we able to pick

---

[9] This suggests one way of viewing the rationale for Russell's "no-class' theory: Since conditions do all the work, why not make do with conditions and drop classes altogether? Of course, we are then on the road to the constructibility theory.

out the right objects and the true membership relationship from the possibilities presented to us? And how can we tell if any two of us are talking about the same objects and the same membership relationship?

None of these objections need deter a determined Platonist from accepting the Maddy view of sets. There is nothing logically impossible about the view. Still, these consequences do seem to prompt in most people with whom I have discussed the matter a definite feeling of scepticism. I should think physicists would be especially dubious of Maddy's views. Few would take seriously a theory which postulates an enormous number of entities that are not physically distinguishable from ordinary physical objects—especially, given that no significant empirical evidence is provided in support of this theory. Also, Maddy's views seem to have far-reaching consequences for physical theory: in calculating the amount of force needed to accelerate a particle some specific amount, most physicists assume that the force is applied only to the particle and not to infinitely many other distinct entities that are not mere parts of the particle. If Maddy's theory were to become accepted in physics, the determination of the mass of an object would be rendered much more difficult. How is one to determine how much of the force applied to the particle is really applied to the particle and not to the mixed sets that are in the exact location of the particle. Of course, Maddy could attempt to resolve this difficulty by *decreeing* that no sets have mass. But surely, if there are such things as sets of physical objects, and if these sets have all the properties Maddy attributes to them, then it should be an empirical question whether or not they have mass: it should not be a question that is to be decided by fiat.

Maddy's doctrine that mixed sets have location in space, can be perceived, etc. was developed to deal with another problem to which mathematical Platonism gives rise: How do we obtain knowledge of the existence and properties of the mathematical objects discussed by mathematicians? How do we obtain knowledge, in particular, of the objects of transfinite set theory? In an earlier work,[10] I disputed Gödel's reasoning in support of his belief in the existence of the objects of transfinite set theory. According to my analysis of his

---

[10] "On a Gödelian Thesis Regarding the Existence of Mathematical Objects", which was read at the International Philosophy Conference held in New York City in the Spring of 1976 (sponsored by the Institute for Philosophical Studies). A revised version of this paper became Pt. I of (Gödelian).

reasoning, Gödel addressed himself to the following sort of question: Given that some people (mostly mathematicians) have such-and-such beliefs and such-and-such experiences, is it reasonable for them to infer that objects of this sort truly exist and that they have the properties attributed to them by our set theories? (Gödelian, p. 218). Much of the paper (Gödelian) was an attack on Gödel's reasoning. Maddy has contested my analysis, claiming that people have perceived and believed in sets throughout history (indeed, she went so far as to claim, in (Perception), p. 191, that even prehistoric people perceived and believed in sets). On this basis, she maintained that there is no need for any theoretic justification of our belief in the existence of these sets, any more than there is a need for a theoretic justification of our belief in the existence of physical objects. The suggestion was that, since we can see sets, there is no problem of justifying belief in their existence or in the properties that we attribute to them.

I shall examine Maddy's reasons for making these historical claims, since they form an important part of her defence of set-theoretical realism. In response to Maddy, it should be noted that it is clearly not an *a priori* truth that people have perceived and believed in sets throughout history and indeed throughout a good portion of prehistory: presumably, such a truth (assuming for the sake of argument that there is such a truth) would have to be a contingent and empirical truth. Hence, it can be reasonably demanded of Maddy that she supply us with evidence for such a claim. How does she know what prehistoric people believed in this regard? What grounds does she have? It is striking that she gives no justification for her claim at all. Some of the things she says, however, suggest one way in which she may have come to think it reasonable to attribute belief in the existence of sets to prehistoric people. She emphasizes that these people must have been able to count (Perception, p. 191); so they must have arrived at numerical perceptual beliefs. But such perceptual beliefs are beliefs "that something or other has a number property" (p. 179). Since, according to Maddy, the most reasonable and simplest supposition to make in this regard is that our numerical perceptual beliefs are beliefs about sets (p. 181), we have reason to think that prehistoric people must have believed in sets. So she may have reasoned.

That she reasoned in this way is further suggested by her analysis of how we acquire perceptual beliefs about mixed sets. She considers the case of a person who opens an egg carton, sees three eggs in it, and acquires in the process a complex of perceptual beliefs: "The numerical beliefs are clearly part of this complex of perceptual beliefs because they can influence the others as well as being influenced by them. . . . I further claim that they are beliefs about a set" (pp. 178–9). If that is how Maddy arrived at her belief that prehistoric people believed in the existence of sets, the cogency of her reasoning can be doubted. For there are many ways of analysing cardinality statements that do not require a belief in sets, but that preserve all the usual arithmetical inferences we make. I have given one such in Chapter 5; and for the relatively simple sorts of arithmetical inferences that prehistoric people could be regarded as making, Hartry Field's analysis discussed in Chapter 8 would also do. Besides, there are the analyses of cardinality given by the early Logicists in terms of concepts or propositional functions. So how can Maddy be sure that the people of prehistory did their cardinality reasoning in terms of sets? Even if the supposition mentioned above, namely that numerical perceptual beliefs are beliefs about sets, is the simplest and most reasonable supposition for *us* to make (which I very much doubt), how can we be confident that such a hypothesis was the simplest and most reasonable supposition for prehistoric people to make? And even if such a supposition were the simplest and most reasonable supposition for prehistoric people to make, why should we conclude that prehistoric people did in fact make it? Does Maddy believe that people always make the simplest and most reasonable suppositions?

In fact, why should we even assume that these primitive people had any clear ideas about what cardinality statements are about? After all, highly educated and intelligent mathematicians and philosophers have given, during the last several hundred years, divergent and, in many cases, incoherent, analyses of cardinality statements. In sum, Maddy's claims that people have perceived sets since prehistoric times seems to be based on very questionable reasoning and hence is a highly questionable step in her attempted resolution of the problem of how we obtain knowledge of the objects of transfinite set theory.

There is, however, another way in which Maddy may have arrived at her conclusion that prehistoric people perceived and believed in sets. It has been suggested to me by Ralph Kennedy that Maddy may have reasoned as follows: Prehistoric people had numerical perceptual beliefs. A numerical perceptual belief is a belief that something or other has a number property. We, who are mathematically and logically sophisticated, have good reason to think that the object of a perceptual belief of this sort is in fact a set, and we have arrived at such a conclusion on the basis of an assessment of all the alternative analyses of cardinality that are available. Sets, then, are the objects of certain perceptions of prehistoric people, that is, prehistoric people did perceive sets, on certain occasions, whether they knew it or not.

But Maddy also claimed that prehistoric people *believed in sets*. How, by this line of reasoning, can we arrive at that conclusion? Notice that by the above line of reasoning, if there are in fact such things as witches, then it is quite possible that people who do not believe in witches still perceive them. For example, if one of my neighbours were, in fact, a witch (unbeknown to me), then by the above way of regarding perception, I have, on a number of occasions, seen a witch, even though I do not believe that there are such things as witches. Thus, we are still faced with the question: How did Maddy arrive at her position that prehistoric people believed in sets? Kennedy has suggested that Maddy may be using the expression 'believe in sets' to mean simply: have beliefs that are in fact about sets. Thus, recall that, according to Maddy, numerical perceptual beliefs are beliefs about sets. (Recall, ". . . I further claim that they are beliefs about a set".) So if, as Maddy claims, prehistoric people had numerical perceptual beliefs, and if these beliefs are, in fact, beliefs about sets, then these people had beliefs about sets and, in the above sense, "believed in sets", whether or not they *believed in sets* under the more usual construal of the italicized phrase.

Let us suppose that Maddy did arrive at her conclusions about prehistoric people in the way suggested by Kennedy. Then her conclusions rest upon some rather questionable premisses: (1) A numerical perceptual belief is a belief that something or other has a number property; (2) the simplest and most reasonable hypothesis as to what kind of thing has a number property is the set-theoretical realist's, so a set is the object of a numerical perceptual belief; and (3) if a numerical perceptual belief is acquired just by looking, a set must have been seen.

Regarding (1), it should be emphasized that Maddy provides her readers with no grounds for this claim: she simply asserts that a numerical perceptual belief is a belief that something or other has a number property. But why must this be so? Suppose that someone looks into my office and as a result acquires a numerical perceptual belief. Suppose, in particular, that the person comes to believe that there is exactly one apple on my desk. Must this belief that the person acquires be a belief that something has a number property? Recall that the analysis of cardinality given in Chapter 5 does not require that there *be* anything that has the cardinality property in question in order that the statement 'There is one and only one apple on the desk' be true. Maddy owes us a justification for her claim that a numerical perceptual belief is a belief that something or other has a number property. As for (2), I simply note that the question of what is the most reasonable and simplest hypothesis to accept in this regard is a very complex and difficult one to assess. I doubt that *anyone* has investigated this question in sufficient depth and thoroughness to make such a judgement. Finally, as I argued earlier, Maddy has given us no good reason for accepting (3).

But whether or not one can make a reasonable case for Maddy's conclusion that prehistoric people "perceived" and "believed in" sets (in the senses of these expressions explained above) there is a serious problem with the philosophical use she makes of this conclusion. As Kennedy has emphasized to me, the principal moral Maddy wants to draw from her conclusion that we have always (ever since prehistoric times) perceived and believed in sets is that *there is no need for a theoretic justification of the existence of sets*, the idea being that the status of sets is on a par with the status of physical objects: If we have always perceived and believed in sets, why do we need to justify their existence? However, the fact that we have *perceived* and *believed in* sets (in Maddy's Pickwickian senses of these psychological terms) does not absolve the set-theoretical realist of the burden of providing a theoretic justification for her belief in sets. This can be easily seen by considering the possibility that there are witches. Suppose that, unknown to us, there are genuine witches who live among us and appear to be ordinary human beings. Then, unknown to us, we ordinary people have been *perceiving* and *believing in* witches (in Maddy's senses of these psychological terms) all along. Would this fact then absolve a proponent of the belief in witches of the burden of providing a theoretic justification of her position? Clearly not. For we would not know that we have been perceiving and believing in witches. We would need a theoretic justification of the existence of witches to justify the claim that we have been perceiving and believed in witches.

Maddy takes great pains to argue that the epistemological situation we are in *vis-à-vis* sets is no different from the one we are in *vis-à-vis* physical objects. We humans have always perceived and believed in sets, she argues, just as we have always perceived and believed in physical objects. But it is now abundantly clear that, even if we accept the above analysis according to which we have always "believed" in sets since prehistoric times, there is a significant difference in the two situations. This difference is masked by Maddy's misleading use of the expressions 'believed in' and 'perceived'. She has no good grounds for claiming prehistoric people perceived and believed in sets, *in the ordinary senses of these terms*, any more than I believe in witches.

Finally even supposing that people have always perceived mixed sets, as Maddy claims, how does that solve the problem of how we gain knowledge of the existence and properties of the objects of *transfinite set theory*? For these objects are not supposed to exist in the physical world? The sets built up from the empty set to give us the usual model of ZF set theory, for example, are pure sets which do not have location in physical space and time, even according to Maddy's theory; so there remains the problem of justifying belief in these objects. And how can one justify the belief in the existence of these objects—which no one that I know of claims to be able to see—without providing some sort of theoretic justification. To put it another way, Gödel and Maddy have expressed belief in the existence of these pure sets of classical set theory. How we gain knowledge of these pure sets is not answered by declaring that we can causally interact with mixed sets, any more than the question of how we gain knowledge of pure spirits (or disembodied minds) is answered by declaring that we can causally interact with mixed spirits composed of minds and bodies (that is, people). Imagine someone arguing that there is no problem of how we gain knowledge of pure spirits because we continually interact causally with mixed spirits.

One can summarize my evaluation of Maddy's views as follows:

1. The view has many counter-intuitive and implausible consequences.

2. Many of the crucial doctrines are not justified by either empirical evidence or philosophical argumentation.

3. Even if the controversial doctrines are not contested, neither the problems of reference I sketched in the beginning of this chapter nor the problem of how we obtain knowledge of the existence and properties of the objects of transfinite set theory is solved.

5. Concluding Comments

Maddy is an Involved Platonist who has made a valiant effort to deal with the theoretical problems that her belief in sets engenders. For this, she is to be commended. However, her efforts have not been successful. But do we need to believe in sets in order to make sense of mathematics?

Let us return for one final visit to the society in which everyone is taught that the referents of mathematical terms are astronomical bodies. Consider the following developments: a philosopher, intrigued by the theories put forward by the Involved Believers, begins to wonder why truths about the existence of astronomical bodies are at all relevant to simple everyday reasoning involving notions of cardinality, ordering, structure, and measurable quantities. This philosopher wonders what difference it would make in the applications of mathematics if the astronomical bodies postulated by the mathematicians did not, in fact, exist? Perhaps the literal truth of this system of mathematics is not what makes it such a useful and theoretically fruitful instrument. Perhaps the indispensability of mathematics for scientific theorizing is not due to the fact that this system of mathematics gives valuable information about strange astronomical bodies. Such thoughts prompt the philosopher to devise a system of mathematics in which no assertions about astronomical bodies are ever made: in this system, there are only theorems that assert the possibility of doing something—of constructing open-sentences of various sorts. Surprisingly, all the scientific theorizing that is done within the framework of the traditional (astronomical) mathematics can be carried out, with only minor changes, within the new system of mathematics.

11

Kitcher's Ideal Agents

ANOTHER view of mathematics that should be compared with the constructivistic view of this work is that of Philip Kitcher. Kitcher, too, wishes to find a way of understanding the mathematical assertions of classical mathematics in a non-Platonic way. He, too, believes that there should be some way of understanding mathematics that is consistent with regarding mathematics as a science, that is, as a body of truths. But, as will become obvious in the following, Kitcher's approach is very different from mine.

1. Kitcher's Account of Mathematical Knowledge

The observant reader will have noticed that, apart from some preliminary observations in Chapter 1, I have said almost nothing up to now about mathematical knowledge. My procedure is very different from that of such philosophers of mathematics as Mark Steiner who begins his (Knowledge) with the premiss that the standard theorems of classical arithmetic are known to be true—a premiss that is then used in drawing conclusions about the nature of mathematics. I prefer not to begin with such a premiss, to some extent because I feel there is so much about the concept of knowledge that is unclear or controversial, but also because such a starting-point assumes that arithmetic is a system of truths—a view that has been challenged by Hartry Field and that is in need of careful philosophical investigation.

Philip Kitcher, on the other hand, is willing to take as the starting-point of his recent book (Nature) the thesis that most people know some mathematical truths and that some people know a great many mathematical truths; the stated goal of his book is "to understand how this mathematical knowledge has been obtained" (Nature, p. 3).

Kitcher makes a radical break with the logic-orientated approach to the philosophy of mathematics that is dominant today. There is a tendency, Kitcher believes, for logicians to regard the mathematician's principal activity as being merely the deducing of theorems from various axioms and definitions. Kitcher wants to set philosophy of mathematics on a new course. Instead of taking the construction and development of formal axioms systems as the model of mathematics, Kitcher suggests that we view mathematics as a kind of evolving lineage. Indeed, he calls his view of mathematical knowledge an "evolutionary theory of mathematical knowledge" (Nature, p. 92). Mathematics has a long history of development in which mathematical theory underwent many drastic changes. Kitcher believes that these changes have not occurred fortuitously, and he suggests that we should investigate the history of mathematics in order to extract rational principles that govern the changes that have taken place.

As will become obvious from the following, Kitcher incorporates into his view of mathematical knowledge two ancestral philosophical models: (1) the Kripke–Putnam causal chain picture of reference (which I described briefly in the previous chapter); and (2) Thomas Kuhn's revolutionary model of scientific change (introduced in Kuhn's (Revolutions)).[1] The causal chain model is reflected in Kitcher's suggestion that our present mathematical knowledge can be explained by a chain of prior knowers: the recent end of the chain contains the authorities of the present community and the chain is held to go back in time through a sequence of earlier authorities (Nature, p. 5). It is theorized that this chain can be traced back to the perceptual knowledge of our remote ancestors.

This idea of a chain of prior knowers is tied to Kitcher's psychologistic analysis of knowledge. Kitcher seems to believe that there are only two types of views of knowledge worth considering: the psychologistic and the apsychologistic. *Both types take knowledge to be a kind of true belief.* The distinction between the psychologistic and the apsychologistic views of knowledge is made in terms of the additional condition(s) the views place on knowledge. Psychologistic views distinguish knowledge from mere true belief in terms of the processes which produce belief. According to psychologistic views, knowledge is true belief which results from the appropriate kind of process. Kitcher calls this approach 'psychologistic' because it emphasizes processes that

---

[1] It should be emphasized that Kitcher does not simply incorporate these ancestral views into his own philosophy, but he makes distinctive modifications of the views.

always contain a psychological event, namely the production of a belief. An example of a psychologistic view of knowledge is Alvin Goldman's causal theory of knowledge, according to which (roughly), in order for X to know that P, the fact that P must (partially) cause X's belief that P (see Goldman (Causal)). Another psychologistic view is the one advanced by Robert Nozick in (Philosophical), according to which X knows that P iff X believes that P, P is true, and in addition:

(1) X has come to believe that P in such a way that if P were not the case, then X would not believe P;

and

(2) X has come to believe that P in such a way that if P were the case, then X would believe P.

Apsychologistic views distinguish knowledge from true belief in ways that are independent of the processes which produce beliefs. Thus, some apsychologistic views attempt to distinguish knowledge from true belief by the addition of a condition that requires of the belief in question that it be evidentially supported by the other beliefs that the agent has, where evidential support is taken to be a logical relation that is independent of the causal relations that obtain among the states themselves. Keith Lehrer has put forward an apsychologistic theory in his book (Knowledge). In order for X to know that P, Lehrer requires that X believe that P, that P be true, and in addition that the following condition be satisfied:

X is completely justified in believing that P in a way that does not depend on any false beliefs

where the clause about false beliefs is meant to rule out Gettier-type counterexamples (see Lehrer (Knowledge), p. 21 and also Gettier (Justified)). Lehrer's approach is apsychologistic because the question of whether or not X is completely justified in believing that P is not dependent on the processes that produced X's belief: X is completely justified if his belief is in the appropriate logical relations with his other beliefs.

As can be seen from the above discussion, psychologistic views are *historical*, whereas the apsychologistic ones tend to be concerned with what is the case at some time. Thus, psychologistic theories investigate the causal history of the belief in question, whereas apsychologistic views tend to focus on what is true of the agent in question at the time of belief.

Kitcher opts for a psychologistic view of knowledge, primarily, it would seem, because he believes that the apsychologistic approach is untenable. He believes that no matter what the apsychologistic view may be, he can describe a case in which a person X believes that P, P is true, the apsychologistic condition(s) are satisfied, but X does not know that P because these conditions are met "fortuitously", so that the "logical interconnections required by the proposal are not reflected in psychological connections made by the subject" (Nature, p. 15; see also p. 16). Kitcher suggests that if X has come to believe that P as a result of a suggestion from a witch-doctor, say, or because he dreamt that P, then regardless of the logical interconnections of X's belief set, X does not know that P. "These possible explanations of the formation of belief enable us to challenge any apsychologistic account of knowledge with scenarios in which the putative conditions for knowledge are met but in which, by our intuitive standards, the subject fails to know" (Nature, p. 15).

What then is Kitcher's own theory of knowledge? X knows that P, Kitcher tells us, if, and only if, X believes that P, P is the case, and X's belief that P was produced by a process which is a warrant for X's belief that P (Nature, p. 17), where the term 'warrant' is introduced by Kitcher to stand for those processes that produce knowledge (p. 26). In short, for Kitcher, knowledge is warranted true belief. I should say here that Kitcher sometimes uses the term 'appropriately grounded' in place of 'warranted'. Thus, knowledge can also be described as appropriately grounded true belief.[2]

But when is a process a warrant for a person's belief? Alas, on this important question, Kitcher does not take a stand. He leaves the notion of warrant unanalysed but tells us that he hopes to provide epistemologists attempting to develop a general account of warrant with much new data (in the form of his mathematical examples).[3] The

---

[2] Kitcher also presents in his book an analysis of a priori knowledge; and he argues that mathematical knowledge is not a priori. Although this material forms an important part of his book, it is not directly relevant to my concerns in this chapter. For this reason, I do not take up this topic here.

[3] Kitcher believes that it is entirely appropriate for him to remain neutral on the question of what a warrant is, because (1) there are concrete proposals in the literature for how 'warrant' should be analysed, and (2) it is still early in the development of psychologistic analyses of knowledge so that the correct version may not yet have been given (Personal correspondence.)

historical accounts he gives of mathematical knowledge do show how, according to this view, warrants are passed on from the mathematical authorities of one time to those of another and provide us with some idea of how Kitcher conceives of warrants.

To understand how the Kuhnian model enters this picture, one needs to appreciate the task Kitcher has set himself. Kitcher starts with the proposition that mathematicians know a great many mathematical truths. This requires that the mathematician's mathematical beliefs be appropriately grounded or warranted. So Kitcher sets out to explain how our present-day mathematical beliefs have become appropriately grounded. It is in the course of giving his explanation that Kitcher borrows some ideas from Kuhn, introducing at the same time, the notion of a practice: "One of Kuhn's major insights about scientific change is to view the history of a scientific field as a sequence of practices. I propose to adopt an analogous thesis about mathematical change. I suggest that we focus on the development of *mathematical practice*" (Nature, p. 163). And just as Kuhn rejected the positivist view of scientific change as being powered, for the most part, by the straightforward accumulation of observational data, so also Kitcher rejects the a priorist's view of mathematical change as consisting, for the most part, of the straightforward accumulation of mathematical knowledge through deductions from axioms. Mathematical change, according to Kitcher, comes about through "interpractice transitions" that are rational even when they are not brought about by deductions from axioms.[4]

---

[4] There is, of course, this difference between Kitcher's analysis of the evolution of mathematical practice and Kuhn's analysis of the evolution of scientific practice: Kuhn sees science developing through a succession of radical shifts of scientific practice—major changes in practice that come about by upheavals that are similar to political revolutions. Kitcher sees mathematical practice changing in a much more rational fashion—through rational interpractice transitions. Another difference is this: Kitcher sees warranted mathematical beliefs as being passed on through a succession of mathematical practices that form a chain of practices, terminating in our present mathematical practice. Hence, there is an accumulation of mathematical knowledge, according to the Kitcher model. Kuhn, on the other hand, does not seem to regard the important shifts in scientific practices as leading to more knowledge—at least knowledge about the physical world (see (Revolutions), pp. 170–3, 205–6). Indeed, Kuhn does not even think that the scientist's goal is to arrive at truths about the physical world; rather, "for the scientist . . . the solved technical puzzle is the goal" (Art, p. 343), and each scientific revolution changes the set of technical puzzles to be solved. For a criticism of this aspect of Kuhn's philosophy of science, see Gary Gutting's introduction to (Paradigms), esp. pp. 16–18.

Kitcher suggests that we view a mathematical practice as consisting of five components: a language employed by the mathematicians whose practice it is, a set of statements accepted by those mathematicians, a set of accepted reasonings that are used by these mathematicians to justify many of the statements they accept, a set of questions selected as important, and a set of metamathematical views regarding such things as standards of proof and definition. Accordingly, a significant change in a practice will involve a change in at least one of the five components—which opens the way to viewing a change in mathematical practice as involving more than just the addition to the set of accepted statements.[5]

Notice that a practice contains a set of accepted statements: these are statements that are accepted by the mathematical community as *truths*. These, then, are the mathematical statements that are believed to be true by the mathematical community. Thus, a practice will contain a set of mathematical beliefs which, if they are in fact true and appropriately grounded, will amount to mathematical knowledge.

Kitcher's idea is to explicate the notion of appropriate grounding in terms of the notion of rational interpractice transitions mentioned above, and in this way, to characterize the nature of mathematical knowledge. According to this view, appropriate grounding can be transferred from one practice to another. And it is maintained that all the elements of the set of accepted statements of our present mathematical practice are appropriately grounded if our present practice is the last link in a chain of practices, each link of which has evolved from its predecessor through a rational interpractice transition, so long as the first link of this chain was obtained by such rational interpractice transition from a set of beliefs that were all perceptually warranted (Nature, p. 225).

To complete this account, Kitcher needs first to explain (*a*) how some basic rudimentary mathematical beliefs can be perceptually warranted, and (*b*) how interpractice transitions that are not accomplished by deductions from axioms can be rational; and second, to show how "we can account for the mathematical knowledge we have by tracing a sequence of these interpractice transitions from rudimentary mathematics to current mathematics" (Nature, p. 227).

[5] One curious consequence of this description seems to be that most present-day French and Japanese mathematicians do not share the same mathematical practice, simply because they do not share the same language.

Regarding (*b*), Kitcher's characterization of rational interpractice transitions is admittedly qualitative and imprecise; but some idea of what Kitcher has in mind can be gleaned from his analysis of key examples. Consider the pre-calculus period of renaissance mathematics. During this time, mathematicians lacked any general method of solving certain kinds of questions that had been asked since the time of the Greeks. These include the problems of constructing tangents and normals, computing the lengths of arcs, calculating areas, and finding maxima and minima. Then, as Kitcher analyses the historical evidence, Leibniz devised a basically simple and unified method of attacking these problems: "Leibniz proposes a calculus of difference and a calculus of sums, both of which give directions for moving from a given algebraic equation to a new equation. The relationships found in the new equation then lead to the solution of the geometrical problems" (Nature, p. 234). The modification of the pre-calculus practice brought about by the acceptance of the mathematical methods proposed by Leibniz and developed independently by Newton is, according to Kitcher, a rational interpractice transition. It was rational to accept the new calculus because it provided simple and unified methods for solving a significant class of mathematical questions. "Question-answering and the systematization of previous problem solutions led to the warranted adoption of the Newton-Leibniz calculus" (Nature, p. 270).[6]

Kitcher provides his readers with a fund of examples from the history of mathematics to show that mathematicians are motivated to undertake foundational research not by rarefied epistemological concerns about lack of rigour in their reasoning, but rather by the desire to answer mathematical questions that are recognized to be important (Nature, p. 268). And he sees the reactions of the mathematical community to Frege's foundational work as supporting the above thesis: Frege campaigned for a major revision of the language of mathematics and also for a programme of foundational research, but his arguments were based largely on epistemological considerations. These arguments failed to provoke a response from mathematicians, according to Kitcher, because "Frege's investiga-

[6] Students of Kuhn's (Revolutions) will see a similarity between Kitcher's emphasis upon problem-solving and question-answering, in determining the rationality of interpractice transitions on the one hand, and Kuhn's emphasis upon the "demonstrated ability to set up and solve puzzles" in comparing rival scientific theories on the other hand (see (Revolutions), p. 205).

tions paid no obvious dividends for mathematical research" (p. 269). Kitcher's thesis is that mathematicians attend to foundational issues only when they become convinced that some important *mathematical* problems cannot be solved without a clarification of mathematical concepts and mathematical reasoning. Since the unclarities and confusions that Frege pointed to did not impede mathematical research, they failed to generate much of a reaction from working mathematicians (Nature, p. 269).[7]

I turn now to Kitcher's discussion of how the chain of mathematical practices can be started without presupposing an appropriately grounded mathematical practice. Mathematical knowledge has been thought by some philosophers to pose a problem for a causal theory of knowledge, such as Goldman's which was described earlier, on the assumption that the standard Platonic interpretation of mathematical statements is accurate (see for example Steiner (Knowledge), pp. 111–16). For, to know that some mathematical theorem is true, according to the causal theory, one's belief must have been (partially) caused by the fact stated by the theorem. Since it is generally maintained that mathematical objects do not, and cannot, enter into causal relationships with anything that exists in physical space, the problem arises as to how we can know any mathematical facts. This general line of reasoning seems to raise a problem for Kitcher's account of mathematical knowledge as well as Goldman's because it suggests that we cannot have grounded mathematical beliefs. Kitcher, as we have seen, is sympathetic with the general ideas underlying the causal theory.[8] But he can avoid this difficulty, he believes, because he rejects the standard Platonic interpretation of mathematics. Mathematical statements, according to Kitcher, should not be analysed as being about mathematical objects that do not exist in space and time. (I

---

[7] Although Kitcher does not say so explicitly, the tone of his remarks on this question suggests that he believes mathematicians were *right* not to concern themselves with the foundational issues raised by Frege (see (Nature), pp. 268–70). If this is what he believes, I certainly would disagree. Let us grant that Frege's programme of foundational investigations "paid no obvious dividends for mathematical research". Should mathematicians be praised for being only interested in investigations that have a mathematicial pay-off? Does Kitcher, a very well-rounded philosopher with broad interests, believe that mathematicians ought to restrict their investigations only to those that pay dividends for mathematical research? If so, why?

[8] I do not mean to suggest that Kitcher accepts the causal theory in the form propounded by Goldman in (Causal). But his psychologistic approach has many points in common with the causal approach.

shall return to Kitcher's anti-Platonic interpretation of mathematical statements shortly.)

So how is this chain of practices ever to get started without
presupposing an already existing mathematical practice with
appropriately grounded beliefs? Kitcher believes that certain of our
beliefs about various operations are warranted by ordinary sense
perception. "We observe ourselves and others performing particular
operations of collection, correlation, and so forth, and thereby come
to know that such operations exist. This provides us with
rudimentary knowledge—proto-mathematical knowledge, if you
like" (Nature, p. 117). Now how were we supposed to have gone
from this rudimentary perceptual knowledge to something that
could pass for mathematical knowledge? The answer, in a word, is
by idealization: "I construe arithmetic as an *idealizing theory*: the
relation between arithmetic and the actual operations of human
parallels that between the laws of ideal gases and the actual gases
which exist in our world" (Nature, p. 109). As was described above
in Chapter 5, the statements of arithmetic are analysed by Kitcher to
be about the ideal operations performed by an ideal agent. The
powers of an ideal agent are thus specified in arithmetic by
abstracting from the accidental limitations of our own powers. In
other words, Kitcher maintains that we in effect characterize the
powers of an ideal agent by the arithmetical principles we accept: an
ideal agent is one who performs the operations specified by our
arithmetical principles. "This stipulation is warranted by our
recognition that the ideal operations we attribute to the ideal agent
abstract from the accidental limitations of our own performance"
(p. 118).

As for Kitcher's claim that one can trace a sequence of rational
interpractice transitions from rudimentary mathematics to our
present mathematical practice, it should be noted that Kitcher does
not attempt to produce a detailed and completely convincing case for
this thesis: that would require, he tells us, covering the whole history
of mathematics. Instead, he concentrates on a specific period of the
history, the development of the calculus, and suggests that tracing
the sequence of interpractice transitions in this period should show
the reader how significant transitions in the history of mathematics
can be seen to be rational (Nature, pp. 227–8). The suggestion is that
if one researches other transitions, one will find the same patterns at
work.

2. An Analysis of Kitcher's View: Details and Doubts

KITCHER'S PSYCHOLOGISTIC ANALYSIS OF KNOWLEDGE

Kitcher's idea of how a present belief can be warranted by a long historical process is supported, in part, by his psychologistic analysis of knowledge, so it is appropriate to begin an analysis of his view of mathematical knowledge with an examination of his reasons for accepting the psychologistic approach to knowledge.

Kitcher opts for a psychologistic view of knowledge because he believes that he has a decisive objection against any apsychologistic view of knowledge. It might be responded that even if his objection to the apsychologistic account is conclusive, *one has not been given sufficient grounds for accepting the psychologistic approach.* It should not be assumed that there are only two possibilities in this regard.

To this, Kitcher has responded (in a letter): You either take into account the actual process that produced the person's belief or you do not. If you do, you are a psychologistic epistemologist. If you do not, you are apsychologistic. I believe that this response makes it clear that Kitcher is willing to consider only analyses of knowledge according to which X knows that P iff X believes that P, P is true, and . . . , where the '. . .' is to be filled in with some condition or conditions. I am not convinced that we should restrict ourselves to analyses of this sort. After all, not all epistemologists find such an approach promising. Some find the search for necessary and sufficient conditions for knowledge a hopeless undertaking. Still others want to investigate more complicated sorts of relationships in analysing knowledge. For example, it might be thought that attributions of knowledge involve a contextual element: one should look for conditions which satisfy:

X knows that P relative to context K iff . . .

The point is, Kitcher's argument against the apsychologistic approach does not compel one to adopt the psychologistic approach.

However, let us accept for the sake of argument Kitcher's assumption that there are only two possibilities. Must we then accept the psychologistic view? I have not been persuaded by Kitcher's reasoning that the apsychologistic approach must be rejected, especially in the case of mathematical knowledge (which, after all, is the kind of knowledge under consideration here).

Recall that Kitcher's objection is based on the conviction that one can always describe cases in which a person X does not know that some proposition P is true, even though X believes that P, P is true, and X's belief that P is related to X's other beliefs in such a way that the apsychologistic epistemologist's condition(s) are satisfied. Thus, consider the following example. Tom has a dream in which his next-door neighbour is elected mayor of Berkeley. As a result of this dream Tom comes to believe that his neighbour is mayor of Berkeley. And, for some strange reason (we are to imagine), the other conditions for knowledge required by the apsychologistic epistemologist, whatever they may be, become satisfied. Kitcher believes that in this case, we should see clearly that Tom does not know that his neighbour is mayor of Berkeley. For the process that produced Tom's belief is not of the appropriate sort to produce knowledge. Tom's belief is so irrational that we would not credit Tom with knowing.

But does this line of reasoning provide us with a decisive refutation of the apsychologistic view? In particular, do we have the means of refuting any apsychologistic view of *mathematical knowledge*? Suppose that scientists are someday able to install in a person the ability to tie a bow-tie or ride a bicycle, much as one can install a new program in a computer. Knowing how to ride a bicycle could then be acquired through an operation. Knowing how to ride a bicycle is an ability, and whether or not one has this ability does not depend upon how this ability was acquired. We would not deny that a person knows how to ride a bicycle just because it was acquired through an operation. I wish to suggest that at least some kinds of mathematical knowledge may be closely related to knowing how.

Consider the following partial account of one kind of mathematical knowledge:

> For certain mathematical propositions P, X knows that P if X believes that P, P is true and X is in a state which is such that X can both justify P (in the sense of being able to produce an acceptable proof of P) and also show, by appropriate explanations, that she understands P.[9]

This is an apsychologistic account because the third condition does not presuppose the existence of a certain kind of process that

---

[9] I wish to emphasize that this account of mathematical knowledge is being put forward only to indicate why I do not consider Kitcher's reasoning in support of the psychologicistic view to be decisive. I am not confident that this account is satisfactory.

produces the belief state in question. What is crucial, according to this account, is the existence of the state of mind—not how the state was produced.

Now reconsider Kitcher's method of refuting apsychologistic accounts of knowledge. He imagines that the belief state is produced in some bizarre way (we can suppose that an operation or a knock on the head produces the state of mind). We would all see, he believes, that in such cases the person in question would not know. But is this at all obvious when we are concerned with mathematical knowledge? Imagine, for example, that in the year A.D. 2500, Tom undergoes an operation performed by a team of surgeons. When Tom awakens from surgery, he is convinced of the truth of a certain theorem of analysis that he had never even considered before. Furthermore, from that point on, he is able to prove the theorem, explain its significance thoroughly, and apply it in a perfectly competent way. According to the above account of mathematical knowledge, Tom knows the theorem in question; and the Kitcher reasoning is supposed to show us that this account of knowledge must be incorrect, because, according to this reasoning, we can all see that, in this situation, Tom would not know the theorem in question. But is it clear that we should refuse to allow that Tom knows this theorem? According to my own intuitions, Tom *does know* the theorem.

Interestingly, what I am inclined to say for such mathematical cases is quite different from what I am inclined to say when it is a "matter of fact" that is supposed to be known. Thus, imagine an analogous case in which Tom awakens from surgery believing that his neighbour is mayor of Berkeley. Then even if the belief were true, I would not be inclined to say that Tom knows that his neighbour is mayor of Berkeley. When it comes to such "matters of fact" then, I agree with the psychologistic epistemologist that Tom does not know.[10] What these examples suggest is that at least some kinds of mathematical knowledge need to be treated differently from typical cases of knowledge of "matters of fact": mathematical knowledge may be closely tied to "knowing how". For knowing how is a kind of ability that does not, in general, depend for its existence on how it was produced. (Cf. the examples discussed below, Chapter 12.) In any case, we have some grounds for questioning Kitcher's refutation of

---

[10] See e.g. Nozick (Explanations), where examples of this kind are discussed in detail (pp. 175 ff.).

apsychologistic accounts of mathematical knowledge, and to that extent, we also have grounds for being sceptical of Kitcher's insistence that knowledge must be psychologistic.

ARE THE THEOREMS OF MATHEMATICS KNOWN TO BE TRUE?

Kitcher believes that the standard theorems of classical mathematics are known by mathematicians to be true. And he develops his theory of rational interpractice transitions to explain how mathematicians have obtained this knowledge. Why do I attribute to Kitcher the thesis that the standard theorems of classical mathematics are known to be true? Because the standard theorems of classical mathematics are among the accepted statements of our present mathematical practice; and all the accepted statements of our present mathematical practice are held, by Kitcher, to be warranted. Furthermore, as I shall show shortly, Kitcher maintains that all these standard theorems are true. Since knowledge is analysed by Kitcher to be warranted true belief, it would seem, from Kitcher's views, that all the standard theorems are known to be true. I have also received verification from him personally (by way of a letter) that he does indeed accept the view being attributed to him here. Of course, this attribution fits Kitcher's opening statement:

In this book, I shall provide a theory about mathematical knowledge. I take as my starting point the obvious and uncontroversial thesis that most people know some mathematics and some people know a large amount of mathematics. My goal is to understand how the mathematical knowledge of the ordinary person and of the expert mathematician is obtained (Nature, p. 1).

As I noted above, Kitcher believes that all the standard theorems of classical mathematics are true. I can be more specific: Kitcher believes not only that the standard theorems of classical mathematics are true and known to be true; he believes that these theorems are *vacuously true*! To see why I make this last claim, one needs both to appreciate Kitcher's analogy between arithmetical statements and the physicist's statements about ideal gases and also to understand Kitcher's analysis of such statements as the Boyle–Charles gas law, which he writes in the following way:

[1] $(x)(Gx \rightarrow (px \times vx = k \times tx))$.

He suggests that one way of understanding [1] is as "part of an implicit specification of the properties which an ideal gas must satisfy" (Nature, p. 116). [1] can thus be regarded as true in virtue of the definition of 'ideal gas'. Kitcher does not believe, however, that our knowledge that ideal gases satisfy [1] can be accounted for entirely by noting that [1] follows from the definition of 'ideal gas'. He claims that one must also explain why the definition of 'ideal gas' is reasonable (p. 17)[11] Similarly, Kitcher holds that the laws of arithmetic he specifies implicitly define what an ideal agent is. And our knowledge that these laws are true cannot be explained merely by noting that they follow from the definition of 'ideal agent': we need to see that this implicit definition is reasonable.

Now Kitcher maintains that no ideal gases and no ideal agents in fact exist. It follows, of course, that nothing satisfies the antecedent of the conditional whose closure is [1]. Hence, [1] is vacuously true. Similarly, it is claimed that the statements of arithmetic "turn out to be vacuously true" (Nature, p. 117, n. 18). Presumably, then, the arithmetical laws of Kitcher's arithmetic, when they are expressed more fully than they are in the book, are each in the form of a universal closure of a conditional with an antecedent that says 'x is an operation of an ideal agent'.

According to this analysis, the sentence [1], which is taken to express the Boyle–Charles law, is vacuously true. But notice that

[1*] $(x)(Gx \rightarrow (px \times vx \neq k \times tx))$

is also vacuously true. Furthermore, we know that [1*] is true. For we know that there are no ideal gases. Hence, [1*] can be deduced by elementary logical reasoning from what we know to be true. Similarly, the arithmetical statement

Every natural number has a successor

is vacuously true because no ideal agents exist. But then the sentence

---

[11] I am not convinced that one needs to explain why the definition of 'ideal gas' is reasonable in order to know that [1] is true. It will be seen shortly that one can deduce the truth of [1] from the premiss that there are no ideal gases—a premiss that we know to be true. So, could not a student of elementary logic, who carries out such a simple deduction, know that [1] is true, regardless of whether this student knows that the definition of ideal gas is reasonable? Kitcher would no doubt disagree. But I suspect that many epistemologists would allow that we could know that [1] is true without first determining if the definition of ideal gas is reasonable. However, it should be noted that the conclusions I draw in this chapter do not depend on rejecting Kitcher's requirements for knowledge.

Every natural number has no successor.

is also vacuously true. Furthermore, if we interpreted the above in the way specified by Kitcher, we would know that the latter is true, since we can deduce it by valid logical inference rules from premises known to be true. In short Kitcher's analysis of arithmetic suggests that there would be an enormous number of true sentences of arithmetic that we would know to be true but that are now universally considered to be false. Surely, that is a reason for thinking that Kitcher has neither accurately analysed what the terms in the sentences of arithmetic truly refer to nor correctly captured the content of what a person, who asserts 'Seven plus three is ten', is truly asserting.

The above considerations also give rise to the question 'Why, according to Kitcher, does the truth value of mathematical statements matter?' We get the answer in the following quotation:

One of the primary motivations for treating mathematical statements as having truth values is that, by doing so, we can account for the role which these statements play in our commonsense and scientific investigations. . . . If mathematics is just a sequence of recreational scratchings, then why do the games we engage in prove so useful? When we draw conclusions from a mixture of scientific and mathematical premises, what accounts for our success? Platonism gains its initial plausibility by recognizing that these questions can be answered if we are prepared to return to the idea that mathematical sentences are what they appear to be, to wit, statements with truth values (Nature, p. 105).

But as we have seen from the above, not only are the standard theorems of arithmetic true, according to Kitcher's interpretation, so also are the negations of these theorems. Thus, both '$2+3=5$' and '$2+3\neq 5$' are true under Kitcher's analysis. Indeed, if we analyse arithmetical sentences in the way Kitcher advocates, '$2+3=6$' can also be seen to be true. But if one "uses" this last sentence in applying arithmetic to a simple cardinality problem, one will very probably run into difficulties. And since all of the above arithmetical sentences are analysed to be true, this would suggest that the truth-value of the sentence in question is not what distinguishes the "good" ones from the "bad". In other words, given the above analysis, how can the truth-value of mathematical statements "account for the role which these statements play in our commonsense and scientific investigations"?

One cannot help wondering at this point just what it was that prompted Kitcher to put forward his "vacuously true" analysis, especially since there are passages in the book in which Kitcher seems to be advocating a different analysis of mathematics: in attempting to explain how the idealizations he postulates are related to the actual world, he seems to regard the connective '→' in [1] to be the subjunctive conditional (see (Nature), pp. 120–1). Under this interpretation, [1] states that if something were an ideal gas, then it would behave according to the Charles–Boyle law. So interpreted, the law would not be vacuously true. Similarly, the statements of arithmetic would not turn out to be vacuously true either. So why did Kitcher opt for the material conditional interpretation? The answer seems to be found in the Benacerraf–Resnik doctrine (discussed in Chapter 7) that one ought to provide a uniform semantical account of mathematical and scientific discourse: It is believed that the Tarskian referential account provides an adequate semantical account of scientific discourse and that we ought to give a semantical account of mathematical discourse to match this. Kitcher seems to accept this doctrine and thus is anxious to make his own interpretation conform to its implications (see (Nature), p. 101, n. 1 and Objection 1 on p. 139). Not knowing how to provide a Tarskian semantics for the subjunctive conditional, he opts for the material conditional.[12]

There is a related reason why Kitcher adopted the material conditional interpretation. This is given in his reply to Objection 3 (Nature, p. 140). The objection is that stipulative theories of truth do not introduce any genuine notions of truth at all. Kitcher responds to it by claiming that "the stipulations [of his theory] are construed as fixing the referents of the expressions employed so that the right referential relations obtain." He thus claims that [1] is not true because we stipulate it to be true—rather, [1] is true because there are no ideal gases. "It is wrong to think of stipulational approaches to truth as at odds with referential explanations. The stipulations are better thought of as deepening the referential explanations, showing why the referential relations hold." In short, Kitcher adopted the

---

[12] See also p. 112, where he suggests that the task of developing a non-standard semantical theory "might raise uncomfortable questions about the relationship between arithmetical and nonarithmetical language". I have received further confirmation of this motivation for his adopting the material conditional analysis. In a letter to me, he reacts to the possibility of using a stronger conditional by saying: "That raises my epistemological hackles, for I don't know how we justify such conditionals."

material conditional interpretation so that he could give the above reply.

In the following sections, I shall sometimes write as if I attributed to Kitcher the subjunctive interpretation of his analysis. This is because I find the subjunctive interpretation more reasonable, despite his explicit rejection of such an interpretation. However, most of what I say will hold regardless of how one interprets the '→' of his analysis.

DO PEOPLE BELIEVE WHAT KITCHER IMPLIES THAT THEY DO?

The most unsatisfactory element of Kitcher's view of mathematical knowledge, I believe, is to be found in his analysis of the content of mathematical statements and beliefs. It is one thing to argue that one can interpret the formal sentences of first-order set theory in the way suggested above; it is quite another to claim that the mathematical statements made by ordinary mathematicians and workaday people do have the content attributed to them by Kitcher's analysis. Does the child, who believes that two plus two is four, believe the proposition about the operations of an ideal agent that is given by Kitcher's analysis as the content of that belief? How can it be plausibly maintained that the child has such a belief? After all, the child is not explicitly taught any such thing. None of the elementary textbooks used in any elementary school provides children with instructions about the operations of ideal agents. Teachers are not taught how to instruct children about what collecting operations ideal agents perform. If, despite this lack of instruction, children do form the beliefs about ideal agents that Kitcher's analysis suggests they do, this would be an absolutely amazing fact—something that would demand scientific explanation. It would suggest that humans have an innate tendency to understand mathematical instruction in terms of the operations of an ideal agent. But why should that be? What possible reasons could there be for such a tendency to develop in the species *Homo sapiens*?

Besides, if an innate tendency of this sort were to be present in all humans, one would think that we would all find Kitcher's analysis completely natural and intuitively correct. Indeed, one would expect that an ideal-agent analysis would have been put forward centuries ago. Besides, how is one to explain the fact that almost everyone denies that what she is asserting, in asserting such simple arithmetical facts as that seven plus three is ten, is the rather complex subjunctive fact about ideal agents put forward by Kitcher's analysis? If the ideal-agent interpretation is innate, why should we be completely unaware of the true content of mathematical statements and beliefs?

It might be suggested that, in arguing for the implausibility of Kitcher's analysis of our mathematical beliefs as I did above, I have been presupposing an "internalist view of belief", according to which the content of a person's belief is completely determined by "what is in the person's head". Putnam's famous "twin earth" example is supposed to show that such an internalist view of belief is untenable (see Putnam's (Meaning)). But have I been presupposing such an internalist view of belief? How does rejecting the strictly internalist view of belief lend credibility to Kitcher's account? Let us allow that what a person believes may be a function of both "internal" and "external" factors. It can then be claimed that "external" factors may be cited in explaining how a person came to have a belief with such-and-such a content. Granting all this, how is Kitcher to explain how the child learning that two plus two is four has come to have the belief about the operations of an ideal agent that is given by his analysis? The child never causally interacts with any ideal agent; and no ideal agent is ever explicitly mentioned by the child's teacher. The teacher may even be a Platonist who believes that numbers are abstract objects and that 'Two plus two is four' states that these abstract objects are related in certain ways. How, then, has the child come to have the belief about an ideal agent that Kitcher's analysis implies that she has?

Part and parcel of Kitcher's analysis of our mathematical beliefs is his analysis of mathematical language: he declares that the sentences of our mathematical languages describe the ideal operations of an ideal agent. In other words, Kitcher adopts the sort of hermeneutic position that Burgess described in his (Why). He thus lays himself open to the kinds of criticisms that were raised in that work.

Consider the following response to the above ideas that was made by Kitcher (in correspondence):

I take most of our linguistic usage to be unthinking borrowed reference, and this applies to mathematics with a vengeance. So the real question is: how could anyone have managed to introduce the talk about ideal agents for others to borrow? (Here, an account is given of how this might have come about.)

But I am not arguing that it is inconceivable that some people form their mathematical beliefs in the way described by Kitcher. Rather, I am claiming that it is implausible to suppose that, in general, people form their mathematical beliefs in this way. So the real question is not: How could anyone have managed to introduce talk about ideal agents for others to borrow? It is conceivable that, at some time, some people in the world meant, by their mathematical pronouncements, what Kitcher claims they meant. What I do find implausible is the thesis that ordinary speakers of English throughout history have meant (and believed) what Kitcher claims they have.

I would also like to point out that Kitcher's analysis requires more than "borrowed reference": it implies that what appear to be referring expressions are, in fact, non-referring expressions. More specifically, it implies that such arithmetical terms as '1' and '4' do not refer at all. According to Kitcher, '2+2=4' is supposed to be a conditional statement with the consequent:

if x is a 2-operation and y is a 2-operation, and z is an addition on x and y, then z is a 4-operation.

How children manage to form such beliefs about the operations of an ideal agent cannot be explained merely by appealing to "borrowed reference".

Given the apparent implausibility of Kitcher's analyses, one would like to know why he ever put forward such doctrines. The answer is not hard to find. Recall that Kitcher wants to explain how it is that we have the mathematical knowledge we do have. This requires an explanation of "the *origins* of mathematical knowledge and an account of the growth of mathematical knowledge" (Nature, p. 96). He claims that the process by which we came to have the mathematical knowledge we now have began with "a scattered set of beliefs about manipulations of physical objects" (p. 226). These early beliefs were, he believes, warranted by ordinary sense perception. To explain the process by which mathematical knowledge has developed out of such humble beginnings, Kitcher sets himself the task of providing "an account of the content of mathematical statements, showing how statements with the content which mathematical statements are taken to have can be known on the basis of perception" (p. 96). He tells us that he is going to provide a non-Platonic account of the content of mathematical statements which will both accommodate the worries of anti-Platonists and also do justice to the concerns that inspire Platonism (p. 96). It is clear that Kitcher's ideal agent theory is intended to provide just this account of the content of mathematical statements. This is why he asserts that "the sentences which occur in mathematics books describe ideal mathematical operations (more exactly, the ideal operations of an ideal subject)" (p. 130).

Kitcher's analysis of mathematical statements is intended to be historically accurate: what our mathematical ancestors truly believed, when they believed some mathematical fact, was that an ideal agent performs such-and-such operations. Thus, he tells us that mathematics "consists in a series of specifications of the constructive powers of an ideal subject" (p. 160), and that *at any stage in the history of mathematics*, "mathematical language will contain expressions referring to or qualifying the operations of the ideal subject" (p. 177). Thus, Kitcher's explanation of how we came to have the mathematical knowledge we have requires that the content of our mathematical beliefs be what is given by his ideal-agent analysis of mathematical statements. Supposedly, the historical process which warrants our mathematical knowledge began with ordinary perceptions of actual agents performing various operations of segregating, collecting, and the like. Prehistoric people idealized these operations and formed the concept of an ideal agent who performs these operations. A mathematical practice was developed as a result of such idealizations. Then, through a series of rational interpractice transitions, mathematics evolved into our present practice. If Kitcher were to allow that, at some point in history, people began to think in terms of abstract mathematical objects instead of ideal agents, this would break the chain of warrants and Kitcher could no longer give the sort of explanation he desires. Thus, he insists that his analysis of mathematical statements gives the actual content of our mathematical beliefs.

It can be seen that Kitcher's account is, in at least one respect, similar to Maddy's: both of these philosophers base philosophical claims on their theories of what prehistoric people believed. Maddy asserts that prehistoric people perceived and believed in sets; Kitcher suggests that prehistoric people developed systems of mathematics by abstracting from the accidental limitations of their own performances of certain kinds of operations to arrive at a theory of what operations an ideal agent would perform. My own view is that to base far-reaching philosophical doctrines on such speculations about how our remote ancestors theorized is very risky indeed.

Let us return to a consideration of the *adequacy* of the view under consideration. Take the child in grammar school who knows some number theory, some geometry, and a little real number theory. We are inclined to allow that the child knows that, with regard to the real numbers, multiplication is distributive over addition. According to Kitcher's analysis, what the child knows is some very complex fact about the operations performed by an ideal agent. Now if one asks the child about these operations, she will express ignorance of them. In fact, by all the usual tests and criteria we use to determine belief, she does not believe what, according to Kitcher's theory, she is supposed to believe. Thus, it is reasonable to wonder just what grounds Kitcher has to support his analysis of the content of mathematical beliefs. Surprisingly, he does not provide any evidence from linguistics or psychology that our mathematical beliefs have the features being attributed to them or that the sentences of our mathematical languages have the meaning his analysis requires. It would seem then that his whole case rests upon the contention that his is the only account of the content of mathematical beliefs and assertions that is satisfactory and that provides us with an explanation of how mathematical knowledge is acquired and of how mathematics is applied in science and everyday life. To assess Kitcher's grounds, we need to delve deeper into the details of Kitcher's analysis of mathematical statements.

KITCHER'S THEORY OF ARITHMETIC

Let us begin with an examination of his analysis of arithmetical statements. Kitcher presents us with a first-order formalization of a part of arithmetic which he calls 'Mill Arithmetic'. The vocabulary of this theory consists of the four predicates:

$$\mathbf{M\ U\ S\ A}$$

and the first six axioms of the theory are given as follows:

1. $(x)\mathrm{M}xx$
2. $(x)(y)(\mathrm{M}xy\rightarrow\mathrm{M}yx)$
3. $(x)(y)(z)(\mathrm{M}xy\rightarrow(\mathrm{M}yz\rightarrow\mathrm{M}xz))$
4. $(x)(y)((\mathrm{U}x\ \&\ \mathrm{M}xy)\rightarrow\mathrm{U}y)$
5. $(x)(y)((\mathrm{U}x\ \&\ \mathrm{U}y)\rightarrow\mathrm{M}xy)$

6. $(x)(y)(z)(w)((Sxy \ \& \ Szw \ \& \ Myw) \rightarrow Mxz)$

The quantifiers of this theory range over operations of some sort (to be specified later). The predicates of the theory are interpreted informally as follows:

'M' stands for the predicate 'is matchable to', where an operation is matchable to another if "the objects they segregate can be made to correspond with one another".

'U' stands for the predicate 'is a one-operation', where a one-operation is performed when a single object is "segregated".

'S' stands for the predicate 'is a successor of', where an operation is a successor of another when the former segregates not only all the objects segregated by the latter, but also a single extra object.

'A' stands for the predicate 'is an addition on'. What is an addition? Suppose A and B are segregative operations, which segregate distinct groups of objects (that is, no object segregated by A is an object segregated by B). Then an addition on A and B combines the objects from both segregations. The important point to notice here is that addition is only performed when one is combining the objects from two segregative operations that are performed on disjoint groups of objects. (This is made perfectly clear by definition (32) on p. 136 of (Nature).)

Interpreting the theory in the above manner, we can see that the first three axioms tell us that the relation 'is matchable to' is an equivalence relation; and axiom (4) asserts that any segregative operation to which a unit operation is matchable is also a unit operation.

The axioms of Mill Arithmetic are supposed to stipulate what operations an ideal agent performs. Consider now the following axiom of Mill Arithmetic:

14. $(x)(\exists y)Syx.$

This tells us that for every segregative operation x that an ideal agent performs, she performs another that is a successor to x. (14) is thus an *axiom of infinity*. Kitcher believes that there are decisive reasons for denying that (14) is true when it is regarded as an assertion about the operations we humans perform—after all, we cannot perform an infinite number of segregative operations. But he declares that the axioms of Mill Arithmetic should be taken as stipulations about the operations that an ideal agent performs; and, he says, "we are justified in introducing the stipulations I take to constitute Mill Arithmetic on the grounds that these stipulations idealize our actual collective activities" (Nature, p. 118). However, it should now be realized that the stipulations of Mill Arithmetic do more than just idealize our collective activity; for if there are only finitely many objects in the universe, then there is a finite limit to the segregative operations that even an ideal agent can perform. Not even an ideal agent can perform an operation that is a successor to a segregative operation x if there are no objects that were not already segregated by x. Thus, Mill Arithmetic not only idealizes the capacities of agents, but it also attributes to the universe an infinity of things to be segregated. Given that it is not known whether there are infinitely many "objects" in the universe, one might question Kitcher's claim that the axioms of Mill Arithmetic are known to be true.

Let us now turn to the addition axiom:

15. $(x)(y)(\exists z)Azxy$.

This axiom stipulates that an ideal agent performs an addition on any segregative operations x and y. But this renders the stipulations of Mill Arithmetic incoherent, for not even an ideal agent can perform the impossible. That (15) requires an ideal agent to perform the impossible can be seen from the fact that x and y need not segregate disjoint groups of objects. Indeed, y might be identical to x.

There are various ways of revising (15) to remove this incoherence. Consider the suggestion made by Craig Bach: suppose that instead of (15), we use:

(15*) $(x)(y)(\exists z)(Mxz \,\&\, (\exists w)Awzy)$.

Now, does this express what is wanted?

A standard theorem of number theorem is:

(#) For all natural numbers x and y, there is a natural number z which is the sum of x and y.

And one would like to know just what fact about ideal agents (#) is supposed to express. Is it what is expressed by (15*)? Another possibility is the following alternative:

$(x)(y)(-(\exists z)Azxy \rightarrow (Eu)(Ev)(Ew)(Mux \,\&\, Mvy \,\&\, Awuv))$.

Notice that the logical form of this alternative is very different from that of (15*). Actually, it is not difficult to think up many other possibilities. It is striking that Kitcher provides us with no general method of translating arbitrary sentences of arithmetic into his Mill Arithmetic.

This suggests the following *ad hominum*: Kitcher believes that a Platonist, who believes that arithmetical statements are actually statements about sets, is faced with a fatal problem. What are the natural numbers? "Apparently, our ancestors discussed them for generations, and, on the view under present discussion, they were talking about sets. But *which* sets?" (Nature, p. 104). One cannot say, claims Kitcher, without making an arbitrary choice; and this gives us strong grounds for rejecting the Platonic view under consideration. But is there not an analogous problem for Kitcher's own view? For Kitcher analyses arithmetical statements in such a way that the child who knows (#) knows some fact about the operations of an ideal agent. But *which* fact? We seem to have another *"embarras de richesses"*.

Now it might be thought that I am merely pointing out a well-known fact, namely that there are many ways of formalizing natural language sentences. (This was suggested to me by Kitcher in correspondence.) Thus, it could be argued that the multiplicity of translation into formal languages does not generate the same sort of problems that the multiple reduction problem raises for the Platonist. But the doubts I am raising are not due to our inability to make a motivated selection of some sentence of Mill Arithmetic as the translation of (#). According to Kitcher, *the true logical form* of (#) is not revealed by its surface form. But what is the true logical form of (#)? It would seem that we cannot say. There are too many options that are equally good. Thus, if, as seems to be the case, we cannot pick out (without making an arbitrary choice) just what fact about the operations of an ideal agent (#) expresses, then a doubt arises as to whether or not (#) really does express a statement about the operations of an ideal agent that is postulated by Kitcher's analysis. This point does not depend on the multiplicity of translation into formal languages. We could allow Kitcher the use of natural languages and the problem will remain. (#) is supposed to express some fact about ideal agents. But what fact? What is its true logical form?

As I mentioned above, Kitcher supplies us with no general method of translating sentences of arithmetic into the sentences of Mill Arithmetic. Interestingly, he does sketch some rules for translating sentences of ZF set theory into sentences of his version of set theory.

KITCHER'S VERSION OF SET THEORY

Kitcher's set theory is formalized in a first-order language with a vocabulary consisting of 'C' and 'E'. 'C' is taken to mean 'is a collecting on', where Cxy iff y is collected by the collecting operation x. "Collectings" are very much like segregative operations, but they are not restricted to collecting only objects—they can also collect collectings. The idea is to have the operational analogues of sets of sets.

'E' is specified to mean: 'is equivalent to', where collectings are equivalent if they collect exactly the same objects. It is important to realize that a person can perform a collecting on objects at one time, and then at a later time perform a collecting on the very same objects. She will have performed two distinct, but equivalent, collectings (Nature, p. 126). Similarly, if someone else performs a collecting on these same objects, she will necessarily have performed a different collecting. Collectings differ in this important respect from collections. However, Kitcher thinks he can translate the usual statements of ZF set theory into the language of his theory by replacing '$x \epsilon y$' with 'Cxy', and '$x = y$' with 'Exy'. Using this method of translation, the Axiom of Extensionality of ZF becomes

$$(x)(y)((z)(Cxz \longleftrightarrow Cyz) \rightarrow Exy)$$

and the Axiom of Unions becomes

$$(z)(\exists y)(x)(Cyx \longleftrightarrow (\exists w)(Cwx \ \& \ Czw)).$$

Kitcher believes that the translations we get in this way from the axioms of ZF can be regarded as true by stipulation—their truth flowing from stipulations about what operations an ideal agent performs. In short, the axioms are thought to be consequences of the conditions we impose on ideal agents (see (Nature), p. 134).

The underlying idea of Kitcher's theory of "collectings" is obtained by translating the informal presentation of the iterative generation of sets to be found in George Boolos's (Iterative) into an informal description of the collectings of Kitcher's ideal agent. We are to view the ideal agent's life as divided into a sequence of stages, in which she performs her collectings. Instead of saying that, at the $(n + 1)$ stage, all collections of objects and collections formed at stages up to and including the nth stage are formed, one says that, at the $(n + 1)$ stage of her life, an ideal agent performs all collectings on all objects and collectings performed up to and including the nth stage. And when it is said that the ideal agent performs all collectings on all objects and the collectings performed earlier, it is meant that the agent performs all *possible* collectings, that is, all collectings that it is possible to perform (see (Nature), pp. 134 and 146).

Kitcher's translations of the axioms of ZF are thought to be inferable from the informal description of the iterative conception obtained. Unfortunately, this is not so. Recall that 'E' is to stand for equivalence and that a statement of the identity of sets '$x = y$' gets translated into a statement of the equivalence of collectings 'Exy'. Since, in set theory, one can infer

$$\ldots x \ldots \longleftrightarrow \ldots y \ldots$$

from '$x = y$', we get as an axiom of Kitcher's theory

19. $(x)(y)(Exy \rightarrow (\ldots x \ldots \longleftrightarrow \ldots y \ldots))$

for any open sentence $\ldots x \ldots$ of the language in question. Kitcher makes no attempt to justify or derive (19) from the principles of the iterative conception of collectings. Had he done so, he would have come to see that (19) is false. To see this, imagine that x and y are distinct but equivalent collectings. Since they are distinct, it is possible that some collecting z collects x but not y. Recalling that an ideal agent performs all possible collectings, it follows that an ideal agent must perform such a z. Thus, we would have

$$Exy \ \& \ Czx \ \& \ -Czy$$

which contradicts (19).

The above point can be made in another way. Kitcher wants to use his theory of collectings to develop a version of Mill Arithmetic. Hence, he needs to define within his version of set theory the predicate 'is a number operation'. This he does (Nature, pp. 136–7) using the identity relation. We can infer that Kitcher's set theory is formalized in the predicate calculus with identity. Suppose, as above, that x and y are distinct, but equivalent, collectings. Then (19) allows us to infer, from 'Exy' and '$x = z$', the identity '$y = z$'. Hence, (19) allows us to infer that if x and y are equivalent collectings, then $x = y$. This runs counter to the basic conception of collectings underlying this theory, according to which there are distinct, but equivalent, collectings.

Kitcher might attempt to deal with this problem in a variety of ways. He might define a new relation 'C^*' by means of the formula:

$$C^*xy \longleftrightarrow (\exists z)(Ezy \ \& \ Cxz).$$

Then, instead of replacing '$x \epsilon y$' with 'Cyx' to translate set theory into Kitcher's theory, use '$C*yx$'.[13]

This suggestion shows how one could obviate the above problem if the problem with (19) arose only because of the predicate 'C'. Unfortunately, as we saw above, the identity relation also raises a problem for the principle. Of course, Kitcher could simply remove identity from his set theory; but if the theory is to be applied in science, it would seem that identity would have to be retained.

But it is not only identity that raises a problem for this solution. Kitcher derives the axioms of his ideal agent version of set theory from more fundamental axioms. These axioms are obtained by translating the axioms of Boolos's *stage theory*, which it is thought capture the iterative conception of set. Thus, one of Boolos's axioms states that every member of a set is formed at an earlier stage than the one at which the set is formed. This gets translated into:

$$(x)(y)(s)(t)((Cxy \ \& \ Oxs \ \& \ Oyt) \rightarrow Bts).$$

It can be seen that Kitcher's version of set theory involves quantification over stages as well as operations, where the sequence of stages is regarded as making up "the collective life of the ideal subject" (*Nature*, p. 132). Notice that the vocabulary of this formal theory includes the predicates 'O' and 'B', where 'Oxs' can be translated by 'x is formed at stage s' and 'Bts' can be translated by 't is before s'. Thus, from the fact that collecting a is equivalent to collecting b, one can infer, by the questionable axiom (19), that if a was formed at stage s, then b was also formed at stage s. But such a conclusion would clearly be false for many cases.

A more fundamental objection can be raised to Kitcher's analysis of set theory. An ideal agent is supposed to perform collectings on physical objects by physically segregating these objects just as we humans collect hazel-nuts by gathering them together. But to get a theory sufficiently powerful to yield a version of set theory, Kitcher needs to have his ideal agents perform collectings on collectings. But how are they supposed to do this? One cannot physically gather together collectings in the way that nuts are gathered. Kitcher suggests that "*when we perform higher-order collectings, representations achieved in previous collecting may be used as materials out of which a new representation is generated*" (*Nature*, p. 129). In other words, when we perform a collecting, we represent this collecting by

[13] This was suggested to me by Klaus Strelau.

means of some symbol; then we collect prior collectings by performing a collecting on the symbols we used to represent the prior collectings. At any rate, this is how Kitcher explains the process.

But the ideal agent is supposed to perform uncountably many such collectings (corresponding to the set of all sets of natural numbers). In order to collect all these collectings in a higher-order collecting, each of these uncountably many collectings must be represented by a symbol. Where does the ideal agent get an uncountable totality of symbols to represent these collectings? And how does the agent assign a distinct symbol to each of these collectings? Do we have any real conception of how this can be done? Furthermore, how does the agent perform *all possible collectings* of the members of this uncountable totality? This ideal agent appears to be more godlike than human. My sense of wonder increases when I contemplate the extremes to which Kitcher is willing to carry his idealizations. For, in order to capture the full richness of the Zermelo–Fraenkel system, he is forced to go beyond the idea that the stages at which the ideal agent is supposed to carry out her collectings correspond to instants in the life of the agent: he thus stipulates that the ideal "subject's activity is carried out in a medium *analogous* to time, but far richer than time" (Nature, p. 146). We thus are to imagine a super-agent performing super-operations in super-time. Many would find the idea of this sort of non-existent being simply incomprehensible.

This takes us to a consideration of

KITCHER'S EXPLANATION OF HOW MATHEMATICS IS APPLIED

Kitcher claims that his analysis of mathematical statements explains how mathematics is applied: "The Millian approach I favor has the advantage over standard nominalist programs that it offers a direct explanation of the applicability of arithmetic. Arithmetical truths are useful because they describe operations which we can perform on any object" (Nature, p. 115). Of course, arithmetical truths are not supposed to give us an explicit description of the operations we can perform: as we have seen, arithmetical truths tell us about the operations of a fictional ideal agent. However, as the previous quotation shows, it is implied that we use the theory of ideal agents to draw inferences about what real agents do, in the way physicists use the kinetic theory of ideal gases to draw inferences about what real gases do. "The relation between arithmetical statements and our actual operations is like that between the laws of ideal gases and actual gases" (p. 116; see also p. 138). The physicist abstracts from certain features of actual gases, in various situations we face, to obtain the idea of an ideal gas. The ideal gas behaves the way actual gases would behave if certain complicating factors were removed. The laws governing the behaviour of ideal gases allows us to deduce, up to varying degrees of approximation, the behaviour of real gases. Similarly, it is suggested, we can use the theorems of arithmetic, to deduce the behaviour of real agents: "we can regard operations which we perform (for example, operations of physical segregation as well as less crude operations) as if they conformed to the notions of the arithmetical operations as they are defined" (p. 121). Thus, let us imagine a possible world in which operations are performed that collectively satisfy the axioms of Mill Arithmetic. Call such a world an M-world. Then, according to Kitcher, "Mill Arithmetic is applicable to our world not because our world is an M-world whose arithmetical operations are our physical operations of segregation, but because, if accidental, complicating factors were removed, our world *would be* such a world" (p. 121).

There is something strange about Kitcher's idea that the usefulness and applicability of mathematics in everyday life or science lies in its property of allowing us to draw inferences about the behaviour of real agents in the way that the kinetic theory of gases allows us to draw inferences about real gases. For one thing, when the physicist uses the kinetic theory to draw inferences about the behaviour of real gases, she makes use of mathematics. It would seem, then, that we would need to use mathematics to apply mathematics, that is, to draw inferences about the behaviour of real agents in the way suggested by Kitcher's analysis. This would be grotesque. On the other hand, if, somehow, we do not need to use mathematics to make such inferences, as Kitcher believes (personal communication), one would like to know why not, since the differences between ideal and actual agents are not at all easy to figure out.

The doubts being expressed here are closely related to other doubts I have about Kitcher's analysis of how mathematics is applied. The big advantage Kitcher claims for his account of how mathematics is applied is that it offers a *direct explanation* of the usefulness and applicability of mathematics. There is no problem in seeing why ideal gas theory is useful: it enables scientists to infer various facts about the behaviour of real gases. Similarly, according to Kitcher's analysis, ideal agent theory (mathematics) enables us to infer facts about the behaviour of real agents. But is set theory concerned with how real agents actually collect things? I would have thought that, if anything, set theory would tell us something about what would be *correct* matching, collecting, and sorting—not what actual agents do. Or consider the case of arithmetic. It tells us that if someone correctly counted the number of fleas in carton A and correctly counted the number of fleas in carton B, thus arriving at the numbers 437 and 7,128 respectively, and at the same time, someone else correctly counted the number of fleas in both cartons, then this other person will have counted 7,565 fleas. Arithmetic does not tell us what numbers actual people will end up with in such situations.

Why do we need to infer how real agents collect and order things in real situations in order to apply mathematics? Why should using mathematics in an engineering problem require that we obtain information about how real agents behave? It was shown in Chapters 5 and 6 how applications of the constructibility theory do not depend upon obtaining detailed information about the behaviour of real agents. So, clearly, systems of mathematics can be applied in a way that does not depend upon drawing inferences about real agents. We thus need grounds for thinking that mathematics is in fact applied in the way described above. But Kitcher has not supplied us with such grounds.

Furthermore, if mathematics is applied in the way sketched by Kitcher, one would think that engineers and scientists would have to learn just how ideal agents operate and how real agents differ from ideal ones. After all, if one is to apply the kinetic theory of ideal gas particles to draw inferences about the behaviour of real gases, then one has to learn what ideal gas particles are and how they behave, and one has to learn how real gas molecules differ from ideal gas particles. But in fact scientists seem to receive no education about ideal agents and how they differ from real ones. Kitcher might reply that learning mathematics just consists in learning how ideal agents operate. But even if this were so, it would surely be highly implausible to suppose that the learning of mathematics consists in learning how the operations of real agents differ from those of an ideal agent.

One final point: suppose that a social scientist wished to construct a useful and enlightening theory of how people in the United States actually collect, match, and segregate things. And suppose that, to achieve this goal, she devises a theory of how an ideal agent does these things. Imagine that this theory attributes to the ideal agent the ability to do an uncountable number of algebraic operations in an instant. It would be surprising if such a theory turned out to be more enlightening to the social scientist than a more realistic theory that described agents with abilities that are closer to those possessed by actual people.

There are other reasons for questioning the idea that the ideal agent theory functions in the way the ideal gas theory does. In the case of the ideal gas theory, we can state precisely in what the idealization consists: namely point masses, perfectly elastic collisions, and no mutual attraction between molecules. Because of this, physicists were able to formulate improved versions of the gas law, such as van der Waal's equation, which yield more accurate predictions. We seem to have no such precise characterizations of the idealization made in obtaining the ideal agent theory, which would enable us to obtain better predictions about our own behaviour in the way the gas laws do for the behaviour of actual gases.[14]

HAS KITCHER ACCOUNTED FOR OUR MATHEMATICAL KNOWLEDGE?

Suppose that a particular mathematical practice M has come about as the result of a sequence of rational interpractice transitions starting from some directly warranted mathematical practice. Kitcher maintains that M will itself be appropriately grounded. We can conclude, from this and Kitcher's theory, that the accepted statements of this grounded practice that are true are warranted (and hence known to be true). But why should we accept such a theory? Why, in particular, should we accept Kitcher's assumption that if a mathematical practice M_2 is obtained from a grounded mathematical practice M_1 by means of a rational interpractice transition, then all the accepted statements of M_2 will be grounded (in the sense in which grounding is sufficient to make a true belief into a case of knowledge)? *Such a principle needs to be justified* (and not simply assumed).

To reinforce this point, consider some examples of rational interpractice transitions that Kitcher discusses. In particular, consider the interpractice transitions to the calculus to which Kitcher

---

[14] The main ideas in this paragraph were put forward in a seminar by Piers Rawling.

devotes so much attention. Basically, Kitcher describes the mathematical works of Newton and Leibniz as providing "algorithms" for answering certain kinds of fundamental questions in mathematics. "The staggering complexity of the techniques devised by earlier mathematics (and by Newton in some of his first papers) thus gave way to a simple and unified method" (Nature, p. 231). Given that the new techniques *worked* (that is, gave results that did not seriously conflict with other more established techniques and that resulted in scientific inferences that were scientifically acceptable), this analysis provides us with grounds for thinking that it was reasonable for mathematicians to adopt the new techniques. And we can understand why Leibniz would advocate the adoption of these techniques, while allowing that the reasoning was not rigorous. We can understand too why he would maintain that the difficulties in understanding how the techniques worked should not stand in the way of using them. But to allow that it was reasonable to accept some mathematical proposition or law on the basis of these new techniques is not tantamount to granting that these pioneering mathematicians *knew* that these propositions or laws *were true*. To show that they were warranted in accepting these propositions or laws, one would have to do more than to show that it was reasonable to accept them.

Kitcher suggests that, in order to assess the rationality of some interpractice transition, the problem-solving benefits of the transition should be weighed against the costs involved in making the change (Nature, p. 199). And he notes that many of the interpractice transitions he describes introduced new methods or concepts that involved significant costs. For example, he allows that the eighteenth-century techniques of finding the sum of infinite series sometimes yielded recognizably false results (p. 199). Nevertheless, he argues, it was rational to adopt the new methods:

Mathematicians, like natural scientists and like everyday people, are sometimes justified in using imperfect tools. Just as it is sometimes reasonable to adopt a solution to an everyday problem which is recognized as having certain deficiencies (or a scientific theory which faces certain anomalies) and to trust that, in time, one will be able to iron out the wrinkles, so it is reasonable for mathematicians to accept proposals for extending a practice which bring costs as well as problem-solving benefits (p. 200).

I can agree that it is frequently reasonable (and rational) to use "imperfect tools". But what Kitcher needs to show is much more than that it was reasonable to adopt these new methods: he needs to show that the new practices were *grounded*. And this requires showing that, despite the striking imperfections in the new methods that were used to obtain the results—despite the fact that these mathematicians realized that they were relying on methods that sometimes yielded recognizably false results—*ALL the accepted results of using these methods were warranted*. But this Kitcher has not done.

Thus, despite the detailed investigation into the concept of knowledge that is to be found in the first chapter of his book, Kitcher's historical analysis of mathematical knowledge presupposes a remarkably weak set of conditions for knowledge: it is presupposed that if it is rational (or reasonable) for X to accept P, X believes P, and P is true, then X knows P. Kitcher seems to have adopted such weak conditions for knowledge in attempting to account for our present mathematical knowledge. He thought he could explain our mathematical knowledge in terms of a chain of practices leading up to our present practice, in which all the accepted true statements of each practice in this chain would be *known to be true*. This is why he says: "I take current mathematical knowledge in general, and current arithmetical knowledge in particular, to be explained by the transmission of that knowledge from contemporary society to the contemporary individual, and by the transmission of knowledge from one society to its successor" (Nature, p. 119). It could thus be plausibly maintained that warrant has been passed on through this chain of practices up to our present practice, so that today we are warranted in believing the standard theorems of our mathematics. But how is one to show that warrant was passed on to the next link in this chain of practices? The task is daunting indeed. By weakening the conditions required for knowledge, the task Kitcher set himself became much more manageable.

It was pointed out earlier that Kitcher does not give any enlightening characterization of the processes that warrant beliefs. Despite this lacuna, Kitcher is sure that if an agent has arrived at her belief in some proposition P in such a way that it is reasonable or rational for her to believe P, then she is warranted in believing P. Should such conditions for knowledge be questioned? Here is the sort of example that has convinced many epistemologists of the dubiousness of taking the above as sufficient conditions for knowledge:

John goes for a ride in the countryside and sees what he takes to be a red barn in location L. But unknown to John, there are, in this region, many expertly made papier-mâché facsimiles of barns. However, given what John knows and has experienced in his lifetime, it may be quite reasonable for him to believe that what he saw was a barn. Let us suppose that, in fact, the object was a barn. We thus can conclude, using the sort of conditions Kitcher makes use of, that John knows that there is a barn in location L. But, as most epistemologists see it, John does not know.[15]

The above criticisms of Kitcher's view of mathematics do not diminish the value of his studies in the history of mathematics. Despite the objections I have given, I am sympathetic to his idea that philosophers of mathematics have much to learn from the history of mathematics. Thus, one of my objections to Field's view of mathematics (given in Chapter 8) rests on certain historical facts. And I am inclined to think that Field would not have put forward his deflationist account of mathematical knowledge (to be discussed in the next chapter) had he paid more attention to the history of mathematics.

ZERMELO'S AXIOMS

I have been questioning Kitcher's general method of accounting for our current mathematical knowledge. Are there reasons for questioning any of Kitcher's specific accounts of how knowledge of some mathematical laws or principles were arrived at? Consider Kitcher's analysis of how we obtained knowledge of the axioms of set theory. It should be emphasized that, for Kitcher, the axioms of Zermelo's set theory are known to be true. He attempts to explain how we obtained this knowledge in the following way: just as it is rational to accept scientific theories on the basis of their power to unify and systematize a large body of data or observed phenomena, so also it is rational to accept a system of mathematical axioms on the basis of the analogous power to unify and systematize a large body of statements already accepted (see (Nature), pp. 218–20). Since Zermelo's system achieved such a unification and systematization—one is able to derive within the system, essentially all of classical analysis—it is rational to accept the axioms of this system. As it was noted above, Kitcher maintains that, if it is rational to accept some mathematical proposition, then one is warranted in believing it. So, assuming that the axioms of

[15] See Nozick (Explanations, pp. 174–5) for a discussion of such examples.

Zermelo's system are true, mathematicians know that these axioms are true.

There are many reasons why I am sceptical of this reasoning. Recall that there are many different axiomatizations that yield such a unification and systematization of classical analysis. Now when there are many competing scientific theories that unify and systematize a large body of data, one does not accept them all: one needs special reasons for accepting one of these theories as the true one. So one would think that special reasons would be needed for concluding, from the above considerations, that the axioms of Zermelo's system are true. Such special reasons are not given by Kitcher.

My own position is very different. I do not commit myself to the position that all the theorems of classical mathematics, literally construed, are known to be true, as does Kitcher. My position regarding how mathematical sentences should be analysed is also very different from Kitcher's. In developing the constructibility versions of cardinality theory and analysis, I did not claim to be giving an analysis of what ordinary mathematical sentences mean. It is not crucial, for my purposes that one determine the literal meaning of mathematical sentences. Unlike Kitcher, it seems to me quite possible that the standard logical interpretation of what mathematical statements mean may be essentially correct. But if something like the logical interpretation, according to which the existential theorems of mathematics assert the existence of abstract mathematical objects, is correct, then I would not want to attribute to any one knowledge of the truth of such existential mathematical statements. For I would need to be given strong reasons for believing that such statements are true—something that has yet to be supplied, so far as I can see. Thus, the mere fact that one can achieve a systematization of all the accepted theorems of classical analysis does not constitute, for me, good evidence for the truth of the axioms.

The above discussion may make it appear to the reader that my attitude towards the question of whether the standard theorems of classical mathematics are true, as well as the question of whether we have knowledge of these theorems, is remarkably close to the position attributed to Hartry Field in Chapter 8—a position, it will be recalled, I criticized in great detail. So it is time to turn to an examination of Field's views on mathematical knowledge.

12

Deflationism and Mathematical Truth

1. Field's Deflationism

ACCORDING to Field's nominalistic view of mathematics (discussed in Chapter 8), mathematics consists, for the most part, of false statements. This gives rise to the following puzzle. Since what are generally thought to be mathematical facts are not, according to this theory, facts at all, what is mathematical knowledge? It cannot be knowledge of mathematical facts. To put the puzzle somewhat differently, what does the mathematician know that non-mathematicians do not know? It is not mathematical facts, since the theorems of mathematics do not state facts. It is in response to this puzzle that Field puts forward a position that he calls "deflationism": roughly, deflationism characterizes mathematical knowledge as knowledge of a purely logical sort.

What knowledge does someone who knows a lot of mathematics possess that those who do not know much mathematics lack? There is, first of all, empirical knowledge. Mathematicians know a great deal about other mathematicians. They know, for example, what mathematical propositions mathematicians accept; what they take to be axioms of set theory, number theory, and the like; how they define mathematical terms; etc. (Is, p. 512). Apart from this sort of empirical knowledge (which I shall ignore in what follows), what separates mathematicians from non-mathematicians, according to the deflationist, is their logical knowledge.

Deflationism differs from Logicism in so far as it does not reduce mathematical assertions to purely logical ones. As we have seen, Field does not analyse mathematical theorems into logical assertions. But he does analyse mathematical knowledge into logical knowledge. For the most part, mathematical knowledge is just knowledge that particular mathematical statements follow logically from certain others. Thus, when a mathematician proves a theorem, she does not learn some new fact about abstract objects according to the deflationist, but she does discover that some statement follows from certain others. Here again, we see a similarity between Field's views and those of Putnam's "If–thenism". But the deflationist does not claim (as did the "If-thenist") that mathematical knowledge is restricted to the kind of knowledge stressed by "If–thenism", namely that something follows from something: mathematical knowledge may involve other kinds of logical knowledge as well. For example, mathematical knowledge may involve knowing that some statement is logically consistent.

Because mathematics is held to be a theory filled with falsehoods, it is a major problem for Field to justify his use of mathematics in science and logic. We saw in Chapter 8 how he attempted to deal with this problem as it pertains to physics. In the Appendix, I examine Field's attempt to justify his use of mathematics in logic. My assessment is that neither of these attempts is truly successful.

But there are other reasons for concluding that Field's deflationist account of mathematical knowledge is unsatisfactory. What knowledge, according to Field, does the child acquire in her mathematics classes? Just that certain statements follow logically from certain other statements and that certain statements are logically consistent. I believe that this view will seem highly implausible to anyone who has taught children mathematics. Most children beginning elementary arithmetic have no real conception of logical consequence. Furthermore, children are not generally taught, in their arithmetic classes, that certain arithmetical statements follow logically from certain other arithmetical statements. So how did Field get himself into the position of defending such a strange view?

We have seen that Field is led to this view of mathematical knowledge by his radical solution to the problem of existence assertions in mathematics. Field accepts the logicist's analysis of mathematical language: indeed, Field writes as if mathematics consisted of theories stated in the formal language of first- or higher-order mathematical logic. Thus, the existence assertions of mathematics get analysed in terms of the standard extensional existential quantifiers of the predicate logic. To avoid the Platonic consequences of accepting the truth of mathematical theorems, Field hit upon the strategy of simply denying the truth of mathematics.

But children are not taught mathematics as axiomatized theories formalized in the predicate calculus. What children are taught very early in their mathematical classes are operations and procedures:

they are taught how to count, add, subtract, multiply, divide, and solve all sorts of arithmetical and algebraic problems. Thus, one of the first things mathematical that a child learns is how to answer "How many?" questions; and children learn very quickly a simple method, namely counting, for determining how many. In learning to count, a child learns such things as that fourteen comes before nineteen. Children also learn various facts about the particular arithmetic operations and calculating procedures that they are taught. For example, children learn that if one correctly performs the operation of adding seven and three, one will get ten. They learn that multiplying (correctly) 47 by 3 yields 141. They also learn a whole host of principles governing these mathematical operations and calculating procedures. For example, they learn that in performing the operation of multiplication the order in which one performs the multiplication does not affect the answer. Multiplying 47 by 3 yields the same number as multiplying 3 by 47. I see no reason for denying that such statements as the preceding one are mathematical or that such statements are true.[1]

In short, I believe that the knowledge possessed by a child who has learned a lot in her mathematics classes differs significantly from that possessed by her schoolmates who have learned little. First, the former will know a great many facts about the results one obtains in applying certain mathematical operations and in following various calculation procedures—things that the latter will not know; and second, the former, but not the latter, will know a large number of facts about what procedures and operations one can use to solve various problems, both practical and theoretical.

I have been suggesting that a significant amount of mathematical education consists in teaching students all sorts of calculating procedures, mathematical techniques, and operations. But how did mathematicians develop these procedures, techniques, and operations? Reading Field's works might give one the idea that mathematicians came upon their techniques and procedures simply as a result of making logical deductions. However, the history of mathematics presents us with a very different picture. Some of the mathematical techniques and procedures that are taught today were devised in ancient civilizations to facilitate the practical tasks of

---

[1] For Field, mathematics seems to be essentially some formalized set theory, such as ZFU, in which case '$3 \times 47 = 141$' gets analysed as a set-theoretical statement that involves reference to sets—a sentence that is held to be false by the deflationist.

estimating and comparing lengths, areas, volumes, etc.[2] And it is hard to believe that these techniques came about simply as the result of some artisan making logical deductions. Furthermore, as was indicated in Chapter 8, some of the techniques of the calculus were developed as a result of extrapolation and analogical reasoning from other techniques and procedures that had proven successful (some of which may have been discovered empirically); and that still others were adopted as a result of vague and unrigorous reasoning we would now find completely unacceptable.[3] Of course, most of those mathematical procedures, operations, and techniques that have stood the test of time can be given a kind of logical justification which will be discussed below. And even in those cases in which mathematical methods were introduced with very questionable supporting reasoning, there were frequently real tests of whether it was reasonable to adopt the methods—tests which came down, in the end, to whether or not one got acceptable results when one used the methods.[4] After all, as was pointed out in Chapter 8, the fundamental theorem of integral calculus was used in determining the areas under

---

[2] See O. A. W. Dilke (Measurement).

[3] Thus, Morris Kline has written: "Because of the difficulties in the very foundation of the calculus, conflicts and debates on the soundness of the whole subject were prolonged. Many contemporaries of Newton, among them Michel Rolle, who contributed a now famous theorem, taught that the calculus was a collection of ingenious fallacies. Colin Maclaurin, after whom another famous theorem was named, decided that he would found the calculus properly, and he published a book on the subject in 1742. The book was undoubtedly profound but also unintelligible. One hundred years after the time of Newton and Leibniz, Joseph Louis Lagrange, one of the greatest mathematicians of all times, still believed that the calculus was unsound and gave correct results only because errors were offsetting each other. He, too, formulated his own foundation for the calculus, but it was incorrect. Near the end of the eighteenth century D'Alembert had to advise students of the calculus to keep on with their study: faith would eventually come to them" (Physical, p. 432).

[4] Cf. J. Grabiner's comment in (Cauchy, pp. 16–17): "Many of the results obtained in the calculus had immediate physical applications; this circumstance made attention to rigor less vital, since a test for the truth of the conclusions already existed—an empirical test." Kitcher's discussion of the development of the calculus in (Nature, ch. 10), also supports the above point. See especially his discussion of Leibniz's version of the calculus on pp. 234–7. Cf. also J. Singh's comment: "The methods of calculus were accepted, not because their reasoning was logically impeccable, but because they 'worked'—that is, led to useful results" (Great, p. 52). M. Kline makes a similar point when he writes: "Not able fully to clarify concepts and justify operations, both men relied on the fecundity of their methods and the coherence of their results and pushed ahead with vigor but without rigor. Leibniz, less concerned about rigor though more responsive to critics than Newton, felt that the ultimate justification of his procedures lay in their effectiveness" (Loss, p. 140).

curves, and the validity of what the theorem said could be checked by elementary methods of determining areas.

The Logicists sought to provide a kind of logical rationale and justification for our mathematical practices—a kind of rationale and justification that we cannot reasonably suppose the originators of these methods to have clearly seen. Another goal which researchers in the foundations of mathematics have set themselves is to construct a clearly articulated framework within which the bundle of principles, techniques, and results that make up what we sometimes call "classical mathematics" can be justified and developed systematically, rigorously, and deductively. The fact that set theory has worked so well in achieving both the above two goals is undoubtedly one of the biggest sources of its attractiveness. One can see why there are philosophers of mathematics, such as Quine, who advocate regarding mathematics simply as set theory.

As did Quine before him, Field treats mathematics as if it were a formalized axiomatic set theory, such as ZF, which he regards as straightforwardly Platonic. Rejecting Platonism, Field adopts the paradoxical position that essentially all of mathematics is false. Kitcher, on the other hand, begins with the premiss that mathematics, as we now teach it in our universities, is true and, indeed, known to be true. Also rejecting Platonism, he is forced to give mathematical sentences a radical interpretation. What, then is my position? I, too, have rejected the Platonic position. But my view of mathematics differs considerably from those of both Field and Kitcher. On the one hand, I have rejected Kitcher's ideal agent analysis of mathematical sentences and I have not accepted Kitcher's reasoning for his conclusion that the axioms of Zermelo's set theory are true. On the other hand, I have criticized Field's doctrine of the falsity of mathematics. Where does that leave me?

2. A Reconsideration of the Veridicality of Mathematics

Let us reconsider, in the light of the previous discussion, the central idea of the argument presented in Chapter 8 for the veridicality of mathematics. I argued as follows:

> For hundreds of years, mathematicians and scientists have regarded mathematics as a developing body of truths; and they

have constructed their mathematical and scientific theories on the assumption that their mathematical beliefs were for the most part true. Many of these beliefs have been checked and rechecked countless times and in countless ways by both sophisticated and elementary methods. Furthermore, this way of proceeding has yielded remarkable results and has been tremendously successful. We thus have some confirmation of the hypothesis that our mathematical beliefs are, to a significant degree, true.

Is not this reasoning incompatible with the position I have just articulated? Have I not suggested that classical set theory, at least when interpreted in the way advocated by the Literalist of Chapter 1, is a false theory? And does this not suggest that I also reject the truth of classical mathematics? After all, my scepticism regarding the truth of classical set theory (when interpreted in the way advocated by the Literalists) is based upon my scepticism about the existence of sets. This scepticism extends to the existence of other mathematical objects such as numbers and functions. So the question arises: How is this scepticism compatible with my position of Chapter 8 according to which our mathematical beliefs are, for the most part and to a significant degree, true?

The problem I take up here was once forcefully brought home to me by John Burgess (in correspondence), when he suggested that my rejection of mathematical objects commits me to regarding our mathematical theories as "radically false". The idea is this: If one denies that there is such a thing as phlogiston, then is one not completely rejecting the phlogiston theory? And if one denies that there is such a thing as caloric, then is one not claiming, in effect, that the caloric theory is utterly and completely false? Burgess believes that, in a similar fashion, in denying the existence of mathematical objects, I am, in effect, rejecting as completely and utterly false classical mathematics.

The phlogiston example certainly does suggest that I have a serious difficulty here. But let us consider other examples. Take the case of Newtonian physics. According to this scientific theory, absolute space exists. But suppose that a scientist denies the existence of absolute space. Must this scientist reject as utterly and radically false Newtonian physics? Surely not. The rejection of absolute space is compatible with the position that Newtonian physics gives a reasonably accurate description of many aspects of the physical world. With this example in mind, let us consider the following example from the history of mathematics.

Henri Lebesgue once made the wise observation in (Measure), p. 13 that "it makes little difference to the success of arithmetic that the metaphysical notions are obscure". He then went on to suggest that numbers are mere symbols (p. 17). Indeed, he described integers as "material symbols intended to give reports of physical counting experiences", and he characterized all numbers as "merely symbols whose purpose was to give reports of physical observations, geometrically schematized" (p. 81). Still later in this work, he wrote: "We have renounced the distinction between the metaphysical number assigned to a collection and the symbol that represents it . . . and we leave for others the problem of dealing with metaphysical problems that are outside our competence" (p. 128).

After Frege, most philosophers of mathematics find it hard to take seriously the suggestion that numbers are mere symbols. But should they reject as utterly and completely false Lebesgue's work on integrals? Surely not. After all, the essential mathematical ideas to be found in this work do not depend on Lebesgue's view of the nature of numbers. As was indicated above, Lebesgue was all too aware of his own unfamiliarity with the intricacies of metaphysical problems.

When Lebesgue gave his famous definition of the integral in (Development), he contrasted it with that of Riemann's. It is obvious that geometric ideas guided and motivated the construction of his definition, as they did in the case of Riemann's. But whereas Riemann broke up the appropriate interval on the x-axis into subintervals of length less than epsilon, Lebesgue recommended breaking up, with the aid of numbers y_i differing among themselves by less than epsilon, the interval on the y-axis bounded by the lower and upper bound of $f(x)$. Corresponding to the subintervals of length less than epsilon, we will then have sets E_i of numbers x for which

$$y_i < f(x) < y_{i+1}.$$

Then, letting η_i be any number between y_i and y_{i+1}, and letting $m(E_i)$, be what is now called the 'Lebesgue measure' of E_i, we can form the *approximating sum*

$$\sum \eta_i m(E_i)$$

so that the limit of the sums so formed, as epsilon goes to zero, can be taken to be the integral we want (Development, pp. 182–3).

Lebesgue gave (Development, p. 181) the following explanation of his method:

One could say that, according to Riemann's procedure, one tried to add the indivisibles by taking them in the order in which they were furnished by the variation in x, like an unsystematic merchant who counts coins and bills at random in the order in which they come to hand, while we operate like a methodical merchant who says:

> I have $m(E_1)$ pennies which are worth $1.m(E_1)$,
> I have $m(E_2)$ nickels worth $5.m(E_2)$,
> I have $m(E_3)$ dimes worth $10.m(E_3)$, etc.

Altogether then I have

$$S = 1.m(E_1) + 5.m(E_2) + 10.m(E_3) + \cdots$$

Now should we reject this definition and the existence theorem that Lebesgue gave in his paper because we reject Lebesgue's conception of number? Surely, even a Quinian, who believes that numbers are sets, would allow that Lebesgue has given an intelligible and fruitful definition and that his existence theorem is essentially correct. Then can I not also maintain that both the definition and the theorem are essentially correct? From my point of view, there is no such thing as a number or a set. But versions of set theory and the theory of real numbers can be reproduced within the constructibility theory, which will allow one to make the sort of statements needed to define the Lebesgue integral. Furthermore, the essential geometrical ideas that go into Lebesgue's definition—such as breaking up the y-axis instead of the x-axis, forming the sets E_i and the approximating sums, and then taking the limit of these sums as epsilon goes to zero—can all be captured in the theory given in Chapter 6. In this way, I can extract, from Lebesgue's paper, what was novel about his definition of the integral, without committing myself to his metaphysical ideas about numbers. And what is essentially Lebesgue's existence theorem can be proved. Thus, I have grounds for maintaining that what Lebesgue did in his paper was correct and sound, even though I reject his analysis of number.

Undoubtedly, many classical mathematicians had beliefs about the nature of numbers, functions, operations, sets, and the like that give the mathematical assertions they made a decidedly Platonic cast. But I see no compelling reason why I should understand the mathematical theorems that have come down to us through history in exactly the way they were understood by the mathematicians who are credited with proving them. I believe that most of these theorems have a mathematical content that can be abstracted from such a Platonic background. Thus, Newton and Leibniz believed that they had developed a method for calculating the areas under certain types of curves. This method relies on the fundamental theorem of integral calculus. From the point of view of the constructibility theory, the Newton–Leibniz method of calculating areas is a sound one, regardless of what views about the nature of numbers, sets, and functions these mathematical pioneers might have had; and what the fundamental theorem says, as expressed in the constructibility theory of Chapter 6, can be seen to be true. For we can extract, from the works of classical analysts, mathematical ideas that are independent of their metaphysical trappings; and these mathematical ideas can be seen to be perfectly sound, by transporting the essential definitions, procedures, and techniques to a new constructibility setting.

At this point, I would like to return to a question that was raised, but not answered in Chapter 1, Section 2. Set theory, or some roughly equivalent mathematical theory, it was claimed, is indispensable for science. Does this not suggest that set theory succeeds in capturing truths in a way in which the theory of perfect beings does not? Do we not have grounds for believing that set theory is true? My answer, of course, is that, although there is much that is true in what is expressed in set theory, that does not give us good grounds for believing in the existence of sets. I mentioned earlier that set theory performs two important foundational tasks: (1) it provides a rationale and justification for our mathematical practices; and (2) it serves as an overall theory within which classical mathematics (including number theory, algebra, and analysis) can be developed. It is not surprising that there are philosophers, such as Quine, who tend to regard classical mathematics as a collection of truths about sets. It is then a small step to conclude that anyone who denies that sets exist must believe that classical analysis is radically false. (Small wonder then that there are nominalists, such as Field, who declare that essentially all of mathematics is false.) But such a conclusion, I have been arguing, is simply not warranted. The first part of my rebuttal consisted in showing how the two important foundational tasks can be accomplished without presupposing the truth of set theory, *literally construed*: the constructibility theory can be used instead. The second part of my argument is directed at showing how rejecting the existence of sets does not commit one to the Fieldian position that mathematical statements must be regarded as false.

Mathematics would not be all that different today if students were taught to develop mathematics within the framework of Simple Type Theory instead of Zermelo–Fraenkel set theory. Given such a type-theoretical setting, it would be hard to insist that my scepticism about the existence of sets implies the radical falsity of mathematics: my constructibility version of Simple Type Theory is simply too close to the Platonic version to make such a claim plausible.

Thus, it should be evident that, from my point of view, set theory is not at all like the phlogiston theory—that is, completely and utterly false. Of course, I do not accept the literal truth of the axioms of Zermelo–Fraenkel set theory, when the language of the theory is interpreted in the way advocated by mathematical logicians. But it should now be evident that, in rejecting the Platonic views of the Literalists, I can still allow that there is much truth to be found in set theory.

Appendix
Field's Nominalistic Logical Theory

I examine here Field's attempt to justify his use of model theory and proof theory in his theoretical work. As I made clear in Chapter 8, there are grounds for doubting the legitimacy of Field's use of set theory and mathematical logic in his explanation of why mathematics is useful in physics. In (Is), however, Field attempts to justify his use of logic and set theory: in particular, he attempts to show why he can use model theory and proof theory in his theoretical work: it is claimed that a deflationist (and a nominalist) can make use of these Platonic theories because of a kind of conservativeness result that he believes he can prove.

The first question we need to answer to see how Field proceeds in this work is this: How are the basic notions of logical consequence and logical consistency to be explained, given the nominalistic framework Field has laid down? He takes certain modal notions as primitives of his logical theory: they are the notions of logical possibility and logical necessity. The symbols 'Pos' and 'Nec' are to be understood, in what follows, as the primitive modal operator symbols '$\diamond$' and '$\square$' in Field's system. Knowledge of consistency and consequence are then explained as follows:

> If A is the conjunction of all members of a body T of mathematical claims . . . , then instead of saying that we have mathematical knowledge that this theory is consistent, why not simply say
>
> (ii*) we know that $\diamond$ A
>
> . . . [And] instead of saying that the claim B follows from the body of claims whose conjunction is A, why not say
>
> (i*) we know that $\square$ (A→B)
>
> where '$\square$' ('it is logically necessary that') is of course defined as '$-\diamond-$' (Is, pp. 515–16).

In the course of developing his deflationist view of mathematical knowledge, Field sets out to justify the use he made of model theory in his earlier nominalistic writings, and it is this attempted justification that can be regarded as a sort of reply to two of the objections I raised in Chapter 8 (see doubts 5 and 6).

Field sets out to show that he is justified in using classical model theory and proof theory to draw conclusions about logical possibility and logical necessity. He claims that, in general, model theory can be used to make inferences about logical possibility and necessity via instances of the "model-theoretic possibility schema" (MTP).

If there is a model for 'A', then Pos A

and the "modal existence schema" (ME)

If there is no model for 'A', then $-$ Pos A.

Similarly, proof theory can be used to provide us with information about logical possibility and necessity via instances of the "modal soundness schema" (MS)

If '$-$ A' is provable in F, then $-$ Pos A

and the "modal completeness schema" (MC)

If '$-$ A' is not provable in F, then Pos A,

this last principle being restricted by the type of sentence A appearing in the schema (Is, pp. 536–7). According to Field, every instance of either of the above schemata is a theorem of "standard mathematics" (for the appropriate F and A), where "standard mathematics" is understood to include both set theory and substitutional quantification (p. 540). Field claims that the reason proof theory and model theory can be used by the deflationist in drawing conclusions about logical possibility and necessity is the *conservatism* of mathematics. According to Field, standard mathematics is *conservative* over the relevant nominalistic theories about which we wish to draw conclusions involving logical possibility; and mathematics is conservative in the following sense: "[*A*]*ny conclusion about the logical possibility or impossibility of nonmathematical claims that can be reached with mathematics can also be reached without it*" (Is, p. 542, italics mine).

It is surprising that Field describes *conservatism* in this way, especially after having corrected his earlier views about conservatism. Recall that in (Science) he described the conservatism of mathematics over nominalistic theories as *deductive* conservatism. Then, in response to the objections of Shapiro, he declared that he should not have appealed to deductive conservatism at all in stating his position, but only *semantic* conservatism. But now, he seems to have returned to deductive conservatism, for in speaking of what can be "reached" by the deflationist, he clearly is suggesting that the deflationist can infer *without mathematics* anything about the logical possibility of non-mathematical claims that can be inferred using mathematics.

There are several aspects of this rather far-reaching conservation claim that bear noting. First, the statement of the claim is fuzzy and it is never made precise in the article. What does it mean to say that a conclusion "can be reached" without the use of mathematics? That it is logically possible that someone could reach it? That we have a procedure for "reaching" such a conclusion? Or what? Unfortunately, Field never provides us with anything like the relatively precise statement of the conservation theorem of (Science). Nor does he even sketch the deductive system by which the deflationist will be able to "reach" without the use of mathematics any non-mathematical claim about logical possibility that can be "reached" using mathematics.

Second, Field does not provide his readers with anything remotely like a proof of his conservation claim. The most he does is to set forth an argument to support his claim that the nominalist can legitimately use the four-lettered principles to find out about logical possibility (Is, p. 543); but this conclusion is considerably weaker than the above italicized conservation claim. After all, it is one thing to claim that *some* inferences about logical possibility based on model-theoretic and proof-theoretic reasoning are legitimate for the deflationist, and quite another to claim that *any* conclusion about logical possibility arrived at using mathematics can be arrived at by the deflationist without the use of mathematics.

Third, even if the strong conservation claim were justified, we still would not have a justification for many of the uses Field makes of Platonic proof theory. Consider the many ways in which various incompleteness results have entered into Field's reasoning. For example, Field was convinced by Shapiro that the representation theorem he had placed so much emphasis on in (Science) cannot be proved in ZFU^2 for NNP^1. Also, Field accepted Shapiro's proof that ZFU^2 is not deductively conservative over NNP^2. Both of these results were proved using Gödel's Second Incompleteness Theorem. One wants to know just how such reasoning can be made intelligible using the above conservation claim. After all, what is concluded is a fact about provability—not a fact about the logical consistency or inconsistency of some nonmathematical claim. Hence, even granting his strong conservation claim, we are still owed an explanation of how Field can use Platonistic proof theory and model theory *in the way he actually does* in his philosophical papers.

Since Field offers no justification for his strong conservation claim, let us consider the justification he gives for the weaker one. Let us consider his justification of the use of just (MTP). In a footnote (on pp. 540–1), Field gives a sketch of a proof of this principle from the axioms of a metamathematical theory M*. He concludes that for any sentence A of some appropriate object language,

Nec $(M^* \rightarrow$ (there is a model for 'A'$\rightarrow$ Pos A)).

From this, it follows in his modal logic (an S5 system) that

Pos $(M^*$ & there is a model for 'A'$)\rightarrow$ Pos Pos A

from which, Field infers

Pos $(M^*$ & (there is a model for 'A'$))\rightarrow$ Pos A

since

Pos Pos A $\longleftrightarrow$ Pos A

is a theorem of an S5 modal logic. This last result then allows one to infer 'Pos A' from the conjunction of

Nec (M*→(there is a model for 'A'))

with

Pos M*.

Field concludes that the deflationist is free to infer 'Pos A' when she has made a deduction of

There is a model for 'A'

from M*, so long as Pos M* is known. By similar reasoning, Field attempts to justify the other, lettered principles.

Let us see how the above line of reasoning has been thought to justify the kind of conservation theorem used in (Science). Resnik's idea in (Ontology) is to show that, on the assumption that "standard mathematics" (that is, M*) is consistent, we can infer, for a nominalistic theory T and a mathematical theory M (both theories of the respective sort described in Chapter 8)

Pos T→Pos (M + T).

We begin with the reasoning in (Science), which shows that

Nec (M*→any model of T can be extended to one of T + M).

Then it is claimed that one can prove in M* by (ME) that if Pos T, then T has a model, and by (MTP) that if M + T has a model, then Pos (M + T). From this, one can obtain

Nec (M*→(Pos T→Pos (M + T))).

Resnik concludes: "Thus all the nominalist needs to conclude that M is conservative over T is to know that M* is possible" (Ontology, p. 202).

There is one problem with the above reasoning. Principles (ME) and (MTP) are concerned with the logical possibility of the truth of single sentences, not theories. Indeed, the modal operators 'Pos' and 'Nec' have been specified to be operators which apply only to sentences; so if 'T' is the name of a set of sentences or a theory, then 'Pos T' has not even been defined.

It may be thought that Field can circumvent this problem by making use of substitutional quantification: for some purposes, a theory can be identified with its axioms and the set of axioms can in turn be identified with the conjunction of its axioms; so if, as Field claims, substitutional quantification is an acceptable nominalistic device for capturing infinite conjunctions (a claim that has been questioned by Resnik in (Ontology, sect. 5), it would seem that (ME) and (MTP) can easily be extended to cover sentences with substitutional quantification. Thus, if Field could justify versions of the above two-lettered principles in which the sentence referred to in the principle captures the content of the relevant theories by means of substitutional quantification, it would seem that the above difficulty could be circumvented.

In response to this suggestion, it should be pointed out, first of all, that even with substitutional quantification, the strong conservation result described in (Science) would not be justified; for not all theories are axiomatizable. Clearly, the above technique will only work for certain sorts of theories. But more importantly, Field himself seems to rule out this method of obviating the above difficulty. For he declares that the principles (MTP) and (ME) would not, in general, hold when the object language sentence includes substitutional quantification, stating:

I suspect that most instances of (MTP) and (ME) in which the instantiation of 'A' contains substitutional quantifiers would be consequences of standard mathematics, but there would be some exceptions in the case of sufficiently powerful instantiations. (Before this can be investigated one must make some decisions as to just how powerful a version of substitutional quantification one wants to invoke. Presumably one should allow *some* form of "impredicative" substitutional quantification . . .) (Is, p. 540, n. 32).

Reviewing Field's justification for his claim that it is legitimate for him to use model-theoretic and proof-theoretic reasoning, it should be noticed that even if the above reasoning were correct, we still would not have a *proof* that it is legitimate for him to use the four-lettered principles. What the above reasoning shows at best is that it is legitimate for Field to use the four-lettered principles *so long as he knows Pos M**. Evidently, Field believes that he has this needed knowledge. For after writing: "What I claim then is that the knowledge of possibility and impossibility that the platonist attains via the four lettered principles can be attained by the deflationist as long as the deflationist *knows that* $\Diamond Ax_M$ . . ." (Is, p. 542, italics mine);[1] he later concludes: "So it is clear that a deflationist can legitimately use model-theoretic reasoning to find out about possibility and impossibility" (p. 543).

I find it puzzling that Field can be so confident that he *knows* that M* is possible (that is, that M* is logically consistent). After all, M* should not be regarded as a weak predicative or constructive system that some anti-realist philosophers require: M* is a very complex and powerful mathematical theory, with some distinctive non-standard features. For example, as I shall make clear shortly, the language of M* is not standard: besides the usual extensional quantifiers of first-order logic, it contains substitutional quantifiers. Furthermore, the derivational system of M* will include an omega-rule of inference. Such a rule allows inferring a sentence beginning with a substitutional quantifier from an infinite number of sentences of a particular sort. Thus, a derivation, in Field's system, can have infinitely many lines. This in turn seems to result in a proof relation for the system that is not effectively decidable! (I shall return to this striking feature of this system shortly.)

But there are other features of M* that make it non-standard. The four-lettered principles are concerned with the logical possibility or impossibility of sentences of quantified modal logic (Is, p. 540). Since all the instances of the

---

[1] 'Ax_M' stands for his conjunction of the axioms of 'standard mathematics', i.e. M*.

four-lettered schemas are held to be mathematical theorems, that is theorems of M* (p. 541), it is clear that modal operators will occur in the language of M*. In short, the derivational system of M* will have to deal with both substitutional quantification and modal operators. With such non-standard features, it seems misleading to me to call such a system, as Field does, "standard mathematics". M* is hardly standard.

Field claims that to say of any axiomatic theory supplemented with an omega-rule that it is consistent implies that the unsupplemented theory is omega-consistent (Is, p. 542). This claim is not at all obviously true. Using the notion of omega-consistency as it was introduced by Gödel in 1931 (and is explicated today in standard textbooks such as Mendleson's (Intro) and Kleene's (ML), a formalized arithmetical theory A with numerals $0, 0', 0'', \ldots$ is omega-inconsistent if there is some formula $F(x)$ which is such that

$$\vdash_A (\exists x) F(x)$$

and also

$$\vdash_A - F(0)$$
$$\vdash_A - F(0')$$
$$\vdash_A - F(0'')$$
$$\cdot$$
$$\cdot$$
$$\cdot$$

What it would mean to say of just any theory that it is omega-consistent is not at all clear. So let us suppose that, in making his claim about the omega-rule, Field was concerned not with just any theory, but rather with the specific theory M*, and that arithmetic has been developed in M* from its set theory by a series of extensions by definition. To evaluate Field's claim, we need to see just how substitutional quantification is defined for M* and also how the substitutional quantifiers are integrated into the theory with its standard extensional quantifiers and modal operators. We also need to know if the omega-rule is to be restricted to sentences with substitutional quantifiers or whether it also applies to sentences with only extensional quantifiers. And we obviously need a much more precise statement of the omega-rule. In short, we need a formalization of M*.

Recall that Field's reasoning in support of his lettered principles rests on the assumption that it is known that M* is consistent. But how can anyone know this in the absence of any precise specification of the theory? Does it not matter how it is specified? Besides, if M* is to be used to provide the kind of proofs given in (Science), it will have to include a strong impredicative set theory with an inaccessible cardinal axiom. Since there are important researchers in set theory who suspect that these strong set theories are inconsistent, one can be excused for being sceptical about Field's claim to know that M* is consistent, especially given the many vaguely characterized non-standard features of the theory and given that he cannot provide us with a proof or even a plausible argument in support of his consistency claim. Is this not somewhat like a mathematician claiming to know that Fermat's Last Theorem is consistent with the axioms of first-order Peano Arithmetic, even though she cannot give even a plausible argument in support of this proposition? Why should we take such a mathematician seriously?

So far, my objections to Field's justification of his use of model theory and proof theory have all been very general. It is time to examine more closely the details of Field's reasoning. Let us consider the justification that Field offers for his lettered principles. In particular, let us examine Field's proof that (MS) is provable from "standard mathematics". Field begins his proof with the requirement that we pick a derivational system F which has the feature that every sentence inferable from a set of sentences is a logical consequence of this set. He then suggests that we could argue by induction on the length of the derivation that each instance of

(MS′) If 'B' is derivable in F, then Nec B

is true. But this would require that he define in his version of "standard mathematics", M*, some predicate that functions like 'true'—a task he wishes to avoid. This is why he requires instead that M* include a substitutional quantifier, InfConj [A], and an omega-rule of inference in its derivational system, which allows one to infer InfConj [A] from its infinitely many conjuncts. He then asserts (without supplying any justification for his assertion) that, with this version of "standard mathematics", he can give the above inductive proof.

Field claims that this reasoning would provide us with a proof that all instances of (MS) follow "by logic alone" from his "standard mathematical theory" (Is, p. 542). He needs to make this claim in order to conclude

(MS*) Nec $(Ax_{M^*} \rightarrow ('-A'$ is derivable in $F \rightarrow -$ Pos A$)$.

But how can he make such a claim since the suggested proof proceeds by mathematical induction? Evidently, Field does not envisage giving the proof by meta-induction, that is, by induction in the metatheory: he seems to think that the inductive proof would be given in M* itself, writing "the entire proof can be done in . . . number theory if you use Gödel numbering" (Is, p. 540).

Let us imagine carrying out Field's proof in the language of M* using Gödel numbering. Let $B(x,y)$ be a formula of N (the number theory of M*) that represents in N the proof relation of the derivational system F. Then an instance of (MS) can be expressed in M* as follows:

$(\exists x)B(x, \ulcorner -S \urcorner) \rightarrow -$ Pos S

where S is any sentence of M*. But the inductive proof Field sketches requires a general statement expressing the proposition that *for every sentence X which* *has a derivation of length less than n,* $-$ *Pos X.* And it is not at all clear just how one can express that general statement in M* using Gödel numbers and number-theoretic relations. We seem to need here a number-theoretic relation $\mathbf{T}(x)$ that expressed truth in M*—something we do not have.

It might be thought that substitutional quantification will allow Field to express the above italicized sentence within the Gödelized scheme sketched above. But just how substitutional quantification could enable one to do this is not obvious. Certainly, Field has not shown us how. And in the absence of any publication working out the ideas of this supposed proof, a significant scepticism about it can arise.

But there is another problem with this line of reasoning. Let us suppose that an "instance" of (MS) has been proved in M*. Then, supposedly, we can infer:

[1]   Nec $(Ax_{M*}\rightarrow((\exists x)\mathbf{B}(x, \ulcorner-S\urcorner)\rightarrow -Pos\ S))$.

But how can the nominalist use this "instance" of (MS*) in order to conclude from her proof of '$-S$' that $-Pos\ S$? Let us grant, for the sake of argument, that she knows Pos Ax_{M*}. It might be thought that she can reason as follows: She can make use of the fact (which she has proved in the metatheory) that

[2]   '$-S$' is provable in F iff $\vdash_N (\exists x)\mathbf{B}(x, \ulcorner-S\urcorner)$.

Thus, she can infer from her proof of '$-S$' that

[3] $\vdash_N (\exists x)\mathbf{B}(x, \ulcorner-S\urcorner)$

and then that

[4]   Nec $(Ax_{M*}\rightarrow(\exists x)\mathbf{B}(x, \ulcorner-S\urcorner))$.

From [1] and [4], she can conclude that

[5]   Nec $(Ax_{M*}\rightarrow -Pos\ S)$.

Using her modal logic and her knowledge that Pos Ax_{M*}, she can conclude from [5] that $-Pos\ S$.

The trouble with this line of reasoning is that it presupposes mathematics: the proof of [2] is carried out in a metatheory using number theory. To prove [2], the nominalist will have to ascend to a metatheory in which one can theorize about the proof capabilities of N and also correlate primitive symbols, sequences of primitive symbols, etc. with the numerals of N. The complicated reasoning required here to prove [2] in such a metatheory involves the use of many mathematical principles such as mathematical induction.

Might not the deflationist arrive at [4] directly without making use of [2]? This does not seem likely since the formula $(\exists x)\mathbf{B}(x, \ulcorner-S\urcorner)$ will contain such an enormous number of occurrences of primitive symbols as to be beyond the (direct) comprehension of humans. The nominalist will not even be capable of writing the formula down, so it does not seem likely that she will be able to prove [4] directly. Besides, if the deflationist restricts her use of (MS) to cases in which she actually derives $(\exists x)\mathbf{B}(x, \ulcorner - S\urcorner)$ from Ax_{M^*}, she will be able to justify practically nothing, and Field's use of classical model theory will certainly not have been justified. The only route open to the deflationist, it would seem is to attempt to justify [4] by means of metamathematical reasoning—something that will require an appeal to mathematical principles.

We can summarize the above point as follows: One can prove a kind of instance of (MS) in M* using Gödel numbering and number-theoretic relations; but to use this sort of instance to draw inferences about logical possibility in the way Field envisages, one needs to make use of results from mathematics. Here again, we find Field presupposing mathematics in trying to show that the deflationist is justified in using mathematics in her logical and scientific reasoning. It would seem that whenever Field tries to justify his use of mathematics, somewhere in his justification, there is presupposed some mathematical results.

There is another problem with Field's attempt to show that the nominalist (and deflationist) is justified in using the conclusions of standard Platonic proof theory "as devices for finding out about logical possibility". Field claims that "proof theory is used for this purpose via the *modal soundness schema*" (Is, p. 537). And by giving the justification discussed above for this schema, he believes that he is justified in making use of standard Platonic proof theory. But consider the deductive system of Field's own version of "standard mathematics". Can the modal soundness schema be used to validate conclusions about logical possibility, on the basis of constructing derivations in such a deductive system, in the way Field suggests? It would seem that it cannot—not, at least, given the way Field attempts to justify (MS). We saw earlier that the proof relation for Field's system of "standard mathematics" is not effectively decidable. Thus, it is not representable in arithmetic. Hence, the attempted justification discussed above cannot be carried out for such a deductive system. Indeed, the type of justification Field attempts cannot be carried out for any deductive system that does not have an effectively decidable proof relation.

I conclude:

1. Field has not proved that it is legitimate for the deflationist to use the lettered principles.

2. Even granting, for the sake of argument, Field's four-lettered principles, Field has not shown how it is legitimate for him to use classical proof theory and model theory in the way he does in his philosophical papers. Nor has he shown how it is legitimate for the deflationist to use the modal analogues of many standard principles of proof theory and model theory that are not consequences of the lettered principles.

3. Field has not even attempted to justify his stronger conservation view

that *any* claim about the logical possibility or impossibility of non-mathematical claims obtainable with mathematics is also obtainable without mathematics.

Clearly, there is yet much for him to do.

POSTSCRIPT

I recently received a revised version of (Is), which is to appear in a collection of Field's papers. This version arrived too late for me to give here a thorough analysis of the revisions Field has made; but I am convinced that the new version no more succeeds than did the earlier one. It seems to me that many of the objections I raised above to the original version will also apply to the revised one. In any case, I can give a few brief comments to indicate why I am not satisfied with his revised defence of his use of Platonic model theory and proof theory in his philosophy of mathematics.

In the revised version, Field no longer attempts to justify his use of Platonic logic in terms of the above lettered principles. Instead, he advances a new set of principles for the deflationist to employ, namely:

(MTP#) If Nec (NBG→there is a model for 'A') then Pos A.

(MS#) If Nec (NBG→there is a proof of '−A' in F) then −Pos A.

(ME#) If Nec (NBG→there is no model for 'A') then −Pos A.

(MC#) If Nec (NBG→there is no proof of '−A' in F) then Pos A.

Alas, the appeal to these new principles seems to be open to the same sort of objections that I raised above. To see this, let us examine just one of these principles, say (MTP#). To make use of this principle, Field will need to know

[#] Nec (NBG→there is a model for 'A').

Drop from consideration, for now, the necessity operator and concentrate on the conditional. The antecedent is a sentence of first-order logic which expresses the conjunction of the axioms of the set theory NBG. The consequent, however, is a sentence of English. Can such a conditional be a *logical* truth of the appropriate kind which would yield the truth of [#]? Here, it is important to keep in mind the way in which Field interprets the necessity operator in question. Field writes:

> Some philosophers (e.g. Carnap) regard the nonmodal logic to which we are adding '◇' as including not only first-order logic properly so-called, but also "meaning postulates" specifying "logical" relations among predicates. Consequently, a nonmodal sentence such as
>
> (2) ∃x (x is a bachelor & x is married)

would count as *logically* false for Carnap, and as a result

(3) $\Diamond \exists x$ (*x* is a bachelor & *x* is married)

also comes out logically false. But I prefer not to follow Carnap in taking meaning relations among predicates to be part of logic. . . . If one adopts this strategy—and I shall—then (2) is not *logically* false; it is *logically* consistent that there be married bachelors (even though it may not be consistent with meaning postulates that there be married bachelors) (Is, pp. 518–19).

The above conditional, then, cannot be logically necessary, in the above sense of the term. For to prove, logically, the conditional in question, we would need to use definitions. For example, we would need a definition of what it is for there to be a model for a sentence. Besides, how could one prove, using logic alone, a conditional with a first-order sentence as antecedent, and a sentence of English as consequent? It seems likely then that the conditional inside the scope of the necessity operator should not be taken to be the hybrid sentence described above. Evidently, Field is using the expression 'there is a model of '−A'' to stand for the Gödelian sentence of NBG that "expresses" what the English sentence does. So interpreted, the conditional can indeed be logically necessary, in Field's sense of the term, and there can be hope that [#] can be proved in first-order logic after all. That Field has such a conditional in mind is further suggested by his explanation of his four new principles; for he claims that "to find out that A is, or is not, logically consistent, it suffices to *derive* a model-theoretic or proof-theoretic statement from standard mathematics" (italics mine). Since the notion of derivations is the formal syntactic counterpart, in mathematical logic, of the intuitive notion of proof, this suggests that he is thinking of the consequent of the above conditional as being the Gödelian set theoretical counterpart of the English sentence that actually occurs as the consequent. Of course, this way of indicating the conditional he wants fits what he wrote in the original version of (Is), especially when he wrote about a proof of a similar conditional: "the entire proof can be done in . . . number theory" (p. 540).

Taking the conditional in this way, I see other problems. First of all, how is the nominalist to arrive at the knowledge that the conditional is logically true? Not by actually deriving the consequent from the antecedent in first-order logic. We know that it is not reasonable to expect anyone to do that. One cannot even be expected to write down the consequent—it is simply too long and complicated a sentence. Rather, the nominalist is going to have to reason to the truth of the conditional by means of metamathematical reasoning, which will involve the use of various metamathematical theorems that validate the required inferences. But to make use of these theorems, the nominalist will have to depend upon mathematics! Here again, Field seems to be presupposing mathematics in trying to show that he does not need mathematics.

A second (and related) problem is this: How is the nominalist to justify principle (MTP #)? Why should the fact that the conditional in question is a logical truth show that A is logically possible? After all, the conditional is simply a very complicated first-order sentence about sets—a sentence that does not even mention A. Here again, the nominalist is going to have to appeal to coding and metamathematical reasoning to arrive at the required justification.

I conclude that Field has failed to give a convincing justification of his use of classical model theory and proof theory.

BIBLIOGRAPHY

Adams, Ernest (Topology), 'The Naïve Conception of the Topology of the Surface of a Body', in P. Suppes (ed.), *Space, Time and Geometry* (Dordrecht, 1973), 402–24.

Benacerraf, Paul (Numbers), 'What Numbers Could Not Be', *Philosophical Review* 74 (1965), 47–73.

—— (Truth), 'Mathematical Truth', *Journal of Philosophy* 70 (1973), 661–79.

—— and Putnam, H. (eds.) (Readings), *Philosophy of Mathematics: Selected Readings* (Englewood Cliffs, 1964).

Bernays, Paul (CoRFM), 'Comments on Ludwig Wittgenstein's *Remarks on the Foundations of Mathematics*', in *The Philosophy of Wittgenstein*, xi. *The Philosophy of Mathematics*, ed. John Canfield (New York, 1986), 77–98.

Bochner, S. (Role), *The Role of Mathematics in the Rise of Science* (Princeton, 1981).

Boolos, George (Iterative), 'The Iterative Conception of Set', *Journal of Philosophy* 68 (1971), 215–31.

Bos, H. J. M. (Curves), 'On the Representation of Curves in Descartes' *Géométrie*', *Archive for History of Exact Sciences* 24 (1981), 295–338.

—— (Equations), 'Arguments on Motivation in the Rise and Decline of a Mathematical Theory; the "Construction of Equations", 1637–ca.1750', *Archive for History of Exact Sciences* 30 (1984), 331–80.

Bourbaki, Nicholas (Architecture), 'The Architecture of Mathematics', *American Mathematical Monthly* 57 (1950), 221–32.

Burgess, John (Dilemma), 'A Dilemma for the Nominalist', unpublished manuscript.

—— (Synthetic), 'Synthetic Mechanics', *Journal of Philosophical Logic* 13 (1984), 379–95.

—— (Why), 'Why I am Not a Nominalist', *Notre Dame Journal of Formal Logic* 24 (1983), 93–105.

Carrigan, R., Jun., and Trower, W. (Superheavy), 'Superheavy Magnetic Monopoles', *Scientific American* 246 (1982), 106–18.

Cartwright, Nancy (Lie), *How the Laws of Physics Lie* (New York, 1983).

Chihara, Charles (Discovery), 'Mathematical Discovery and Concept Formation', *Philosophical Review* 72 (1963), 17–34.

—— (Gödelian), 'A Gödelian Thesis Regarding Mathematical Objects: Do They Exist? And Can We Perceive Them?' *Philosophical Review* 91 (1982), 211–27.

—— (Ramsey), 'Ramsey's Theory of Types: Suggestions for a Return to Fregean Sources', in D. H. Mellor (ed.), *Prospects for Pragmatism: Essays in Memory of F. P. Ramsey* (Cambridge, 1980), 21–47.

Chihara, Charles (Simple), 'A Simple Type Theory without Platonic Domains', *Journal of Philosophical Logic* 13 (1984), 249–83.

—— (V-C), *Ontology and the Vicious Circle Principle* (Ithaca, 1973).

Churchill, R. (Complex), *Introduction to Complex Variables and Applications* (New York, 1948).

De Gandt, François (Réalité), 'Mathématiques et Réalité Physique au XVII Siècle', in *Penser les Mathématiques* (Paris, 1982), 167–94.

Detlefsen, Michael (Hilbert's), *Hilbert's Program: an Essay on Mathematical Instrumentalism* (Dordrecht, 1986).

Dilke, O. A. W. (Measurement), *Mathematics and Measurement* (Berkeley, 1987).

Field, Hartry (Comments), 'Comments and Criticisms on Conservativeness and Incompleteness', *Journal of Philosophy* 82 (1985), 239–60.

—— (Is), 'Is Mathematical Knowledge Just Logical Knowledge?', *Philosophical Review* 93 (1984), 509–52.

—— (Realism), 'Realism and Anti-Realism about Mathematics', *Philosophical Topics* 13 (1982), 45–69.

—— (Science), *Science Without Numbers* (Princeton, 1980).

Fisher, Charles (Death), 'The Death of a Mathematical Theory: a Study in the Sociology of Knowledge', *Archive for History of Exact Sciences* 3 (1966), 137–59.

Forbes, G. (Modality), *The Metaphysics of Modality* (Oxford, 1985).

Frege, Gottlob (Foundations), *The Foundations of Arithmetic* 2nd edn., trans. by J. L. Austin (Oxford, 1959).

—— (Laws), *The Basic Laws of Arithmetic: Exposition of the System*, trans. and ed., with an introduction by M. Furth (Berkeley, 1964).

Geroch, Robert (Physics), *Mathematical Physics* (Chicago, 1985).

Gettier, D. A. (Justified), 'Is Justified True Belief Knowledge?', *Analysis* 23 (1963), 121–3.

Gillies, D. A. (Frege), *Frege, Dedekind, and Peano on the Foundations of Arithmetic* (Assen, The Netherlands, 1982).

Gochet, Paul (Ascent), *Ascent to Truth: A Critical Examination of Quine's Philosophy* (Munich, 1986).

Gödel, Kurt (Cantor's), 'What is Cantor's Continuum Problem?', in Benacerraf and Putnam (Readings), 258–73.

—— (Russell's), 'Russell's Mathematical Logic', in Benacerraf and Putnam (Readings), 211–32.

Goldman, Alvin (Causal), 'A Causal Theory of Knowing', *Journal of Philosophy* 64 (1967), 357–72.

Gottlieb, Dale (Economy), *Ontological Economy: Substitutional Quantification and Mathematics* (Oxford, 1980).

Grabiner, Judith (Cauchy), *The Origins of Cauchy's Rigorous Calculus* (Cambridge, Mass., 1981).

Guttings, Gary (ed.) (Paradigms), *Paradigms and Revolutions: Appraisals and Applications of Thomas Kuhn's Philosophy of Science* (Notre Dame, 1980).

Hazen, Allen (C Semantics), 'Counterpart-Theoretic Semantics for Modal Logic', *Journal of Philosophy* 76 (1979), 319–38.

Hemple, Carl (Nature), 'On the Nature of Mathematical Truth', in Benacerraf and Putnam (Readings), 366–89.

Heyting, Arend (Int), *Intuitionism: An Introduction* (Amsterdam, 1956).

Joseph, Geoffrey (Many), 'The Many Sciences and the One World', *Journal of Philosophy* 77 (1980), 773–91.

Kaplan, David (Heir Lines), 'Transworld Heir Lines', in Michael Loux (ed.), *The Possible and the Actual* (Ithaca, 1979), 88–109.

Kline, Morris (Loss), *Mathematics: The Loss of Certainty* (Oxford, 1980).

—— (Physical), *Mathematics and the Physical World* (Garden City, 1963).

Kitcher, Philip (Nature), *The Nature of Mathematical Knowledge* (New York, 1984).

—— (Plight), 'The Plight of the Platonist', *Nous* 12 (1978), 119–36.

Kleene, S. C. (ML), *Mathematical Logic* (New York, 1967).

Kripke, Saul (Naming), *Naming and Necessity* (Cambridge, Mass., 1972).

—— (Semantical), 'Semantical Considerations on Modal Logic', in Leonard Linsky (ed.), *Reference and Modality* (London, 1971), 63–72.

Kuhn, Thomas (Art), 'Comment on the Relation of Science and Art', in his *The Essential Tension* (Chicago, 1977), 340–1.

—— (Revolutions), *The Structure of Scientific Revolutions*, 2nd edn. (Chicago, 1970).

Lear, Jonathan (Aristotle's), 'Aristotle's Philosophy of Mathematics', *Philosophical Review* 91 (1982), 161–92.

—— (S and S), 'Sets and Semantics', *Journal of Philosophy* 74 (1977), 86–102.

Lebesgue, Henri (Development), 'The Development of the Integral Concept', Pt. II of (Integral), 178–94.

—— (Integral), *Measure and the Integral*, ed. Kenneth May (San Francisco, London, and Amsterdam, 1966).

—— (Measure), 'Measure of Magnitudes', Pt. I of (Integral), 9–175.

Lehrer, Keith (Knowledge), *Knowledge* (Oxford, 1974).

Lewis, David (Counterpart), 'Counterpart Theory and Quantified Modal Logic', in M. J. Loux (ed.), *The Possible and the Actual* (Ithaca, 1979), 110–28.

Lindsay, Robert and Margenau, Henry (F of P), *Foundations of Physics* (New York, 1963).

Lloyd, Elisabeth (Evolutionary), *The Structure and Confirmation of Evolutionary Theory*. To be published by Greenwood Press.

Maddy, Penelope (Perception), 'Perception and Mathematical Intuition', *Philosophical Review* 89 (1980).

Maddy, Penelope (Reply), 'Set Theoretic Realism: a Reply to Chihara', unpublished MS.

Malament, D. B. (Rev), Review of *Science without Numbers*, *Journal of Philosophy* 79 (1982), 523–34.

Mates, Benson (El), *Elementary Logic*, 2nd edn. (New York, 1972).

Mendleson, Elliott (Intro), *Introduction to Mathematical Logic*, 3rd edn. (Monterey, 1987).

Moss, J. (Sprouts), 'Some B. Russell Sprouts (1903–1908)', in *Conference in Mathematical Logic—London '70* (Berlin, 1972), 211–50.

Nozick, Robert (Philosophical), *Philosophical Explanations* (Cambridge, Mass., 1981).

Nye, Mary Jo (Molecular), *Molecular Reality: A Perspective on the Scientific Work of Jean Perrin* (London, 1972).

Plantinga, Alvin (Transworld), 'Transworld Identity or Worldbound Individuals?', in M. J. Loux (ed.), *The Possible and the Actual* (Ithaca, 1979), 146–65.

Polya, G. (Induction), *Mathematics and Plausible Reasoning*, i. *Induction and Analogy in Mathematics* (Princeton, 1954).

Putnam, Hilary (Meaning), "The Meaning of 'Meaning'", in D. Davidson and G. Harman (eds.), *Language, Mind and Knowledge*, Vol. vii of *Minnesota Studies in the Philosophy of Science*.

—— (MMM), *Mathematics, Matter and Method* (Cambridge, Mass., 1981).

—— (Thesis), 'The Thesis that Mathematics is Logic', in R. Schoenman (ed.), *Bertrand Russell: Philosopher of the Century* (London, 1967), 273–303.

—— (What), 'What is Mathematical Truth?', in his (MMM), 60–78.

Quine, W. V. O. (Carnap), 'Carnap and Logical Truth', in his *The Ways of Paradox and Other Essays* (New York, 1966), 100–25.

—— (Dogmas), 'Two Dogmas of Empiricism', in his *From a Logical Point of View*, 2nd edn. (Cambridge, Mass., 1953), 20–46.

—— (Exist), 'Existence and Quantification', in his *Ontological Relativity and Other Essays* (New York, 1969), 91–113.

—— (Facts), 'Facts of the Matter', in R. W. Shahan and C. Swoyer (eds.), *Essays on the Philosophy of W. V. Quine* (Norman, Okla., 1979), 155–69.

—— (Kaplan), 'Reply to Kaplan', in D. Davidson and J. Hintikka (eds.), *Words and Objections* (Dordrecht, 1969).

—— (NF), 'New Foundations for Mathematical Logic', in his *From a Logical Point of View*, 2nd edn. (Cambridge, Mass., 1961), 80–101.

—— (Posits), 'Posits and Reality', in his *The Ways of Paradox and Other Essays* (New York, 1966), 233–41.

—— (Reification), 'Logic and the Reification of Universals', in his *From a Logical Point of View*, 2nd edn. (Cambridge, Mass., 1961), 102–29.

—— (Scope), 'The Scope and Language of Science' in his *The Ways of Paradox and Other Essays* (New York, 1966), 215–32.

—— (Speak), 'Speaking of Objects', in his *Ontological Relativity and Other Essays* (New York, 1969).

—— (Types), 'On the Theory of Types', *Journal of Symbolic Logic* 3 (1938), 125–39.

—— (W and O), *Word and Object* (Cambridge, Mass., 1960).

—— and Goodman, N. (Steps), 'Steps Toward a Constructive Nominalism', in N. Goodman, *Problems and Projects* (Indianapolis, 1972), 173–98.

—— and Ullian, J. S. (Web), *The Web of Belief* (New York, 1970).

Resnik, Michael (How), 'How Nominalist is Hartry Field's Nominalism?', *Philosophical Studies* 47 (1985), 163–81.

—— (MaSoP:E), 'Mathematics as a Science of Patterns: Epistemology', *Nous* 16 (1982), 95–105.

—— (MaSoP:O), 'Mathematics as a Science of Patterns: Ontology and Reference', *Nous* 15 (1981), 529–50.

—— (Ontology), 'Ontology and Logic: Remarks on Hartry Field's Antiplatonist Philosophy of Mathematics', *History and Philosophy of Logic* 6 (1985), 191–209.

—— (Review), Review of *Science Without Numbers*, *Nous* 27 (1983), 514–19.

Rhees, Rush (Continuity), 'On Continuity: Wittgenstein's Ideas, 1938', in *The Philosophy of Wittgenstein*, xi. *The Philosophy of Mathematics*, ed. John Canfield (New York, 1968), 214–67.

Rooney, Phyllis (PhD), 'Fregean Type Theory: Logical Aspects, Philosophical Implications, Historical Consequences', PhD thesis, Berkeley, 1982.

Schoenfield, Joseph (Logic), *Mathematical Logic* (Reading, Mass., 1967).

Shapiro, Stewart (M and R), 'Mathematics and Reality', *Philosophy of Science* 50 (1983), 523–48.

—— (Conserv), 'Conservativeness and Completeness', *Journal of Philosophy* 81 (1983), 521–31.

Singh, J. (Great), *Great Ideas of Modern Mathematics: Their Nature and Use* (New York, 1959).

Stalnaker, Robert (Inquiry), *Inquiry* (Cambridge, Mass. and London, 1984).

Steiner, Mark (Mathematical), *Mathematical Knowledge* (Ithaca and London, 1975).

Struik, Dirk (Concise), *A Concise History of Mathematics*, 2nd edn. (New York, 1948).

Suppe, Frederick (Structure), *The Structure of Scientific Theories* (Urbana, 1974).

Suppes, Patrick (Scientific), 'What is a Scientific Theory?', in S. Morgenbeser (ed.), *Philosophy of Science Today* (New York, 1967), 55–67.

Thompson, Paul (Perspective), 'The Structure of Evolutionary Theory: A Semantic Perspective' *Studies in History and Philosophy* 14 (1983), 215–29.

van Fraasen, Bas (Image), *The Scientific Image* (New York, 1980).

van Fraasen, Bas (Semantics), 'On the Extension of Beth's Semantics of Physical Theories', *Philosophy of Science* 37 (1970), 325–39.

Wittgenstein, Ludwig (Investigations), *Philosophical Investigations* (New York, 1953).

——(Remarks), *Remarks on the Foundations of Mathematics* (New York, 1956).

INDEX